American Tactical
Advancement
in World War I

American Tactical Advancement in World War I

The New Lessons of Combined Arms and Open Warfare

JEFFREY LaMONICA

McFarland & Company, Inc., Publishers
Jefferson, North Carolina

Photographs and map are from Kenyon Stevenson, *The Official History of the Fifth Division USA: During the Period of Its Organization and of Its Operations in the European World War, 1917–1919* (Washington, D.C.: Society of the Fifth Division, 1919).

LIBRARY OF CONGRESS CATALOGUING-IN-PUBLICATION DATA

Names: LaMonica, Jeffrey, author.
Title: American tactical advancement in World War I : the new lessons of combined arms and open warfare / Jeffrey LaMonica.
Other titles: New lessons of combined arms and open warfare
Description: Jefferson, North Carolina : McFarland & Company, Inc., Publishers, 2017 | Includes bibliographical references and index.
Identifiers: LCCN 2017019416 | ISBN 9781476664194 (softcover : acid free paper) ∞
Subjects: LCSH: United States. Army. American Expeditionary Forces—History. | Unified operations (Military science)—History—20th century. | United States. Army. Infantry—Drill and tactics—History—20th century. | Military doctrine—United States—History—20th century. | Tactics—History—20th century. | World War, 1914–1918—Campaigns—Western Front. | World War, 1914–1918—Trench warfare. | World War, 1914–1918—United States.
Classification: LCC D570 .L36 2018 | DDC 940.4/1273—dc23
LC record available at https://lccn.loc.gov/2017019416

BRITISH LIBRARY CATALOGUING DATA ARE AVAILABLE

ISBN (print) 978-1-4766-6419-4
ISBN (ebook) 978-1-4766-2845-5

© 2017 Jeffrey LaMonica. All rights reserved

No part of this book may be reproduced or transmitted in any form or by any means, electronic or mechanical, including photocopying or recording, or by any information storage and retrieval system, without permission in writing from the publisher.

Front cover image: American troops ride on World War I tanks going forward into battle in the Argonne Forest, France, September 26, 1918 © 2017 Shutterstock

Printed in the United States of America

McFarland & Company, Inc., Publishers
 Box 611, Jefferson, North Carolina 28640
 www.mcfarlandpub.com

Table of Contents

Introduction   1

  I. AEF Combined Arms Doctrine   13

  II. AEF Open Warfare Doctrine   41

  III. AEF Combined Arms and Open Warfare Training   54

  IV. AEF Combined Arms and Open Warfare in Action   88

Conclusion: U.S. Army Combined Arms and Open Warfare After the Great War   122

Chapter Notes   149

Bibliography   159

Index   165

Introduction

The Great War of 1914 to 1918 precipitated a revolution in military affairs, introducing weapons and tactics that changed the course of warfare forever. Modern industrialized warfare forced the belligerents to adapt along a broad learning curve, moving from nineteenth-century linear attack formations to trench warfare and ultimately to semi-mechanized mobile combat. After entering the conflict late, the United States Army raced to catch up with its allies and an enemy who already had three years of experience with this new type of war. The last phase of the Meuse-Argonne Offensive in November 1918 presented the American Expeditionary Forces with a final opportunity to implement the hard lessons learned in six months of combat and survival on the western front. The AEF's attempts to educate and train its soldiers and officers manifested in the tactical planning and combat performance during this last American attack from 1 November until the armistice ten days later. By then, American officers and enlisted men were more experienced and somewhat better trained than at any point prior in war. The AEF's advances in combined arms warfare—the coordinated use of infantry and its weapons, artillery, vehicles, aircraft, communications, and logistics for the purpose of achieving a strategic objective—were apparent during selected episodes in this final operation. Furthermore, the campaign displayed the capacity of some AEF units to employ General John J. Pershing's open warfare tactics. Pershing's

Introduction

vision of fire-and-maneuver as an effective alternative to trench warfare was becoming a real possibility by late 1918. The AEF barely hit its tactical stride, however, when Imperial Germany agreed to a ceasefire on 11 November. Had the war lasted into 1919, large portions of the AEF would have been as tactically advanced as any army on the western front.

This study examines the extent to which the AEF mastered the tactical lessons of the Great War. It also explores some of the factors that limited that progress. The AEF's commander-in-chief, General Pershing, believed that the AEF's tactical versatility would justify its existence as an independent fighting force in Europe. He was determined to break the trench-warfare stalemate and restore mobility to the western front with his own brand of open warfare. By fall 1918, Pershing finally published his open warfare vision, and his subordinates began to understand the importance of combined arms and open warfare in creating a war of movement. This was a noteworthy achievement, considering other belligerents needed at least two years of trial and error on the western front to arrive at the same realization. The British Army took until mid–1916 to publish tactical manuals based on what it had learned since 1914. The French did not revise their doctrine to include new techniques until 1917.

Nevertheless, executing new tactics posed a major challenge to the U.S. Army in the Great War. Although combined arms and open warfare techniques eventually found their way into AEF tactical doctrine, training failed to prepare American units to successfully carry out these innovations. Pershing sacrificed tactical instruction to rush American soldiers to the frontlines after the Imperial German Army broke through the Allied lines in spring 1918. This left most AEF divisions to learn combined arms and open warfare through battlefield experience and the impetus to survive in modern industrialized combat. The final phase of the Meuse-Argonne Offensive was the last American attack of the war, and arguably the most expertly executed. During the last eleven days of the six-week Meuse-Argonne campaign, the AEF demonstrated a much improved combat performance. This was especially true among American units which had gained experience in actual combat in the preceding months. Those eleven days pro-

Introduction

vide an ideal case study for assessing the cumulative results of the tactical lessons learned by the AEF during the Great War.

Most historical treatments of the AEF published before the 1980s do not acknowledge its tactical development. They provide sweeping assessments of American battlefield performance and draw broad conclusions concerning the U.S. Army's general contribution to the defeat of Imperial Germany. More recent scholarship, however, provides a deeper understanding of the United States' impact on the Great War by evaluating all facets of the AEF. The majority of these works dismiss the AEF as tactically stagnant and inept. A few scholars delve deeper into the evidence to reveal and more fairly assess AEF tactical growth, such as Mark E. Grotelueschen's *The AEF Way of War: The American Army and Combat in the First World War*. This study strives to fit in with this trend by exploring the AEF's aptitude in two specific tactical areas, combined arms and open warfare. This work will show how the AEF came to appreciate these techniques by late 1918 and explain why it was not able to execute them with greater success. Ultimately, formal training failed to prepare the AEF for modern industrialized warfare. Survival instinct and combat experience, however, fostered enough tactical learning to enable American divisions to keep gaining ground until the armistice. The U.S. Army would revisit and build upon the AEF's experimentation with combined arms and open warfare during the interwar period and the Second World War. Furthermore, the AEF's brand of learning by fighting continued to be the army's method for tactical growth from 1941 to 1945 and still persists in the twenty-first century.

This work analyzes an array of U.S. Army tactical manuals and pamphlets to uncover the extent to which combined arms and open warfare techniques became a part of published AEF doctrine. It also explores the shortcomings of AEF training, such as Pershing's placement of manpower needs at the front ahead of thorough instruction and the limited availability of weapons for training purposes. Finally, this study demonstrates how the impulse to survive on the battlefield and the demands of modern industrialized warfare ultimately instigated tactical innovation by the fall of 1918. An examination of the Fifth Division from 1 November until the armistice will provide evidence

Introduction

for this growth. This final stage of the Meuse-Argonne Offensive was the last test of battle for the AEF.

The first chapter of this study identifies combined arms techniques in AEF tactical doctrine published between spring 1917 and fall 1918. It reveals references to the coordinated use of infantry, field artillery, chemical agents, machine guns, mortars, grenades, automatic rifles, small arms, and special weapons in a sampling of U.S. Army training manuals and tactical pamphlets. The chapter demonstrates the codification of AEF's combined arms by late 1918.

Combined arms tactics present the challenge of using of various arms in concert to achieve a battlefield goal. Integrating new arms with existing arms further complicates this task. From the battle-chariot charging alongside Ancient Egyptian foot soldiers to the unmanned aerial vehicle (UAV) providing reconnaissance for American infantrymen in Afghanistan, combined arms are an ongoing tactical phenomenon. Combined arms in the Great War appear crude in comparison to those of the twenty-first century. Nevertheless, all the belligerents on the western front realized the importance of using infantry, small arms, artillery, vehicles, aircraft, communications, and logistics in unison by 1918. For the British, French, and Germans, three years of costly combat experimentation proved that combining arms was necessary for traversing the modern industrialized battlefield. The AEF had only six months to learn and implement the same combined arms lesson.

Putting combined arms into practice faltered as a result of poor coordination. American infantry and artillery experienced trouble communicating due to the technological limitations that existed in 1918 and, at times, carelessness. High friendly fire casualties resulted from improperly placed artillery barrages. Inaccurate artillery bombardments left enemy machine-gun positions intact when the infantry moved forward. These artillery support issues were just some of the byproducts of poor communication.

Improvements in combined arms coordination required more time, training, and experience than the tempo of the war allowed. The unprecedented speed of the AEF's advance in November 1918, reaching a pace of five miles a day, presented overwhelming challenges to the immature and under-trained American communication and logistical

Introduction

systems. For example, the AEF advanced beyond its communication lines by the third day of the final phase of the Meuse-Argonne Offensive. Information from the combat zone took up to five hours to reach the U.S. First Army Headquarters in Souilly, France, on 3 November. This delay made it impossible to process tactical intelligence and orchestrate spontaneous operational adjustments. Therefore, as the Imperial German Army fell back, AEF units that made unexpected breakthroughs and territorial gains moved forward with exposed flanks and no artillery cover. Other units missed opportunities to gain ground when they stopped to extend their lines of communication back to their command centers. The AEF had no communication redundancy. Contact was lost when enemy artillery fire severed wires between communication stations. The concept of backup connection lines simply did not exist in 1918. These communication limitations hampered the depth of command at every level of the AEF.

A shortage of horses, motor vehicles, tanks, air support, and aerial reconnaissance plagued U.S. Army operations as well. The Americans could only muster eighteen of the five hundred tanks Pershing requested for the last assault and 740 of the five thousand aircraft he expected at war's end. This left the AEF short of essential combined arms components.

The second chapter highlights evidence of General Pershing's open warfare concept in AEF tactical doctrine. It identifies flexible formations, fire-and-maneuver, infiltration tactics, and battlefield initiative and improvisation in a sampling of U.S. Army publications from spring 1917 to fall 1918.

Pershing's open warfare vision stemmed from his determination to restore mobility to the western front and his confidence in the skill and fighting spirit of American officers and enlisted men. He believed open warfare relied on expert marksmen to provide effective suppressing fire and individual bravery to flank the enemy and close with the bayonet. The commander-in-chief's brand of open warfare represented a combination of traditional tactical principles, such as offensive spirit and hand-to-hand combat, and new trends in battlefield survival, such as infiltration and flanking maneuvers. Pershing's tactical ideas, however, contained flaws. Other belligerents learned earlier in the war that

Introduction

élan and the bayonet counted for little on the modern industrialized battlefield. Although the Japanese launched successful bayonet charges as recently as the Russo-Japanese War in 1905, the volume of machine guns and artillery on the western front in 1914 made it virtually impossible to get close enough to engage the enemy with cold steel. Furthermore, the most effective open warfare tactics relied on support from advanced weapons, motor vehicles, and aircraft. The AEF proved slow to mix open warfare infantry tactics with advanced weapons and other arms.

The fact that a clear definition of Pershing's open warfare concept was not published until the fall of 1918 was highly problematic. On 5 September 1918, just two months before the war's end, the AEF finally received standardized combat instructions calling for irregular infantry formations, flanking enemy strong points, and battlefield improvisation. Despite issuing 54,968 copies of Pershing's *Combat Instructions* to the AEF before the end of the war, there were not enough weeks left before the armistice to afford time to learn and implement these new tactics. Furthermore, large AEF divisions, each with a paper strength of one thousand officers and twenty-seven thousand enlisted men, were difficult to maneuver in the field. An American division contained nearly twice the number of troops as a typical European division. The British and French urged the U.S. Army to form divisions of between sixteen and twenty thousand men, but the commander-in-chief argued in favor of these massive divisions. Pershing believed larger divisions could absorb more casualties and, therefore, remain in combat for longer periods of time. The size of AEF divisions compounded the challenge of conducting open warfare in late 1918, as American commanders struggled to control these huge bodies of manpower on the battlefield.

The third chapter establishes insufficient and rushed training as the primary shortcoming in AEF tactical development. It measures the tactical rhetoric revealed in the previous chapter against descriptions of instruction in a sampling of U.S. Army training manuals, AEF training memorandums, Pershing's cablegrams, general orders, and bulletins, as well as post-war reports and the reflections of AEF commanders. This chapter concludes that hasty and inadequate training

Introduction

failed to fully prepare the AEF to employ combined arms and open warfare. Therefore, the American combined arms and open warfare that occurred successfully in November 1918 came primarily from survival instinct and battlefield experience.

Despite the U.S. Army's efforts to update published doctrine during the Great War, AEF officers and enlisted men did not receive adequate combined arms and open warfare training. In the fall of 1918, Pershing responded to escalating manpower demands at the front by abbreviating all AEF training regimens. Approximately one thousand American soldiers arrived in France each day during the Meuse-Argonne Offensive. The commander-in-chief's goal was to send these troops to the front within several days. This had an adverse impact on the AEF's tactical versatility. The frontal assault became the most common American solution for reducing enemy strong points. Furthermore, the lack of advanced weapons, such as tanks and aircraft, for instructional purposes limited the AEF's ability to practice sophisticated combined arms prior to entering combat. These shortcomings in combat preparation turned the western front into an unforgiving classroom for most American divisions.

The fourth chapter is a case study of the AEF's First Army and its Fifth "Red Diamond" Division during the final phase of the Meuse-Argonne Offensive. The fact that Pershing's open warfare concept remained nebulous until September 1918 had the hidden benefit of allowing AEF divisions several months of combat to improvise, experiment, and engage in the experiential learning typical of the U.S. Army's history. AEF officers and enlisted men devised their own tactical innovations to survive on the western front. The resulting incarnations of open warfare and combined arms often proved more effective than those published by the commander-in-chief and the U.S. War Department. The strategic planning and combat decisions of AEF commanders during the last phase of the Meuse-Argonne Offensive indicated that combined arms and open warfare had permeated the AEF's collective mindset primarily through battlefield experience, not published doctrine or training. Even after the publication of Pershing's *Combat Instructions*, much was left to interpretation on a divisional level and manifested itself in many different ways on the battlefield.

Introduction

General Pershing turned the AEF into an army group by dividing it into two armies on 16 October 1918. Lieutenant General Hunter Liggett's First Army contained three American corps with nineteen divisions. First Army was unorganized, understrength, and low on supplies when Liggett took command. He spent two weeks resting, regrouping, reinforcing, resupplying, and preparing his army for the next major push. Liggett also took this time to cram in some open warfare and combined arms training in infiltration tactics and infantry/artillery coordination. He held meetings to impress his interpretation of Pershing's open warfare on his subordinates. General Liggett insisted that his division commanders bypass enemy strong points and focus on capturing and consolidating immediate objectives.

Liggett called for the combined use of all available arms and resources, including nearly one million soldiers and four thousand artillery pieces, when he launched his 1 November attack. His combined arms plan included artillery support for the infantry, chemical agents, and close air support. Creeping artillery barrages preceded infantry units to their objectives, mustard gas bombardments saturated the German lines, and aircraft bombed enemy positions during the assault. Furthermore, truck-borne infantry pursued the retreating Germans and prevented them from establishing fresh defensive positions. General Liggett's two-week preparation period paid off in logistical improvements as well. First Army efficiently evacuated wounded soldiers from the front, and special "rolling kitchens" kept pace with the advancing army.

The Fifth Division was among the first elements of Liggett's First Army to cross the Meuse River on 3 November. The division received little tactical training before its arrival on the western front. Pershing interrupted the Fifth's training to rush the division to the front. For this reason, evidence of the Fifth Division's use of combined arms and open warfare demonstrates how survival instinct and combat experience necessitated tactical growth.

The Fifth was a Regular Army division organized at Camp Logan, Texas, on 11 December 1917. Despite its Regular Army designation, the division consisted primarily of volunteers from the eastern half of the United States between Michigan and Florida. It contained the Ninth

Introduction

One of Lieutenant General Hunter Liggett's rolling kitchens keeps pace with the Fifth Division's rapid advance during the final days of the Meuse-Argonne Offensive (probably 6 November 1918).

Infantry Brigade, Tenth Infantry Brigade, and Fifth Field Artillery Brigade. The Sixtieth Infantry Regiment, Sixty-First Infantry Regiment, and Fourteenth Machine-Gun Battalion comprised the Ninth Infantry Brigade. The Tenth Infantry Brigade contained the Sixth Infantry Regiment, Eleventh Infantry Regiment, and Fifteenth Machine-Gun Battalion. The Nineteenth Field Artillery Regiment, Twentieth Field Artillery Regiment, Twenty-First Field Artillery Regiment, and Fifth Trench Mortar Battery made up the Fifth Field Artillery Brigade. The Fifth's divisional troops were the Thirteenth Machine-Gun Battalion, Seventh Engineers Regiment, and Ninth Signal Battalion.

After some sporadic stateside training, the Fifth Division arrived in France in spring 1918. The division participated in less than one month of instruction in Bar-sur-Aube, France, before entering the frontline with the French Army. The division participated in the Saint Mihiel Offensive as part of the U.S. First Army's IV Corps. The Fifth then moved to the Meuse-Argonne sector on 16 September and spent twenty-six days on the offensive as part of General Liggett's First Army.

Introduction

Pershing placed Major General Hanson E. Ely in command of the Fifth Division on 17 October 1918. Ely had graduated from the U.S. Military Academy at West Point, New York, in 1891, taught military science at the State University of Iowa from 1897 until 1898, served in the Philippine Insurrection from 1899 until 1901, and observed the Imperial German Army in Europe in 1906. He commanded the Twenty-Eighth Infantry Brigade of the First Division at the Battle of Cantigny and the Third Brigade of the Second Division at Saint Mihiel.

For the final phase of the Meuse-Argonne Offensive, General Pershing charged First Army with capturing the vital railway junction at Sedan, France. The Imperial German Army used the north/south rail line running through Sedan to shuttle ammunition, supplies, and reinforcements along the entire western front. The Fifth Division, totaling 21,128 men, was First Army's leading element and spearhead for its final assault from 1 to 11 November. The division broke the Kriemhilde Line on 1 November and crossed the Meuse River two days later. The Fifth employed combined arms and open warfare during this final push. General Ely utilized extensive engineer, artillery, and machine-gun support as his infantry crossed the Meuse River and captured the heights, woods, and villages beyond. Furthermore, the men of the Fifth Division used infiltration tactics to flank and neutralize German machine-gun positions in the wooded elevations along the Meuse.

By the armistice, the Fifth Division was within artillery range of the last line of German defenses before the Franco-German border. The eleven-day operation cost the division 1,807 casualties. This figure represented 21 percent of the division's total casualties during the war. The Fifth Division played a crucial role in making Pershing's Meuse-Argonne Offensive a strategic victory. In the last eleven days of this battle, the Fifth Division helped to place the AEF in position to invade Germany by 1919. The "Red Diamond" Division learned the tactics that enabled this achievement by fighting and surviving on the western front.

The final chapter asserts that the AEF's strides in combined arms and open warfare served as the basis for U.S. Army tactical development during the interwar years and the Second World War, when experiential learning on the battlefield once again became the army's

Introduction

method of tactical development. Depression-era budget cuts and diplomatic isolationism impeded the army's efforts to fully expand upon the AEF's tactics during the interwar years. The U.S. Army in 1941, therefore, picked up where AEF left off with a new battlefield learning curve during the Second World War. The chapter concludes with a brief exploration of U.S. Army combined arms and open warfare tactics and training in the early twenty-first century and how the army continues "learning by fighting."

I

AEF Combined Arms Doctrine

At the dawn of the twentieth century, tactics for each combat arm of the U.S. Army simply mirrored infantry doctrine. For example, the army's *Field Service Regulations 1905* mentioned combined arms, but did not describe the role of different arms in its execution. Detailed techniques for each combined arms component did not appear in U.S. Army doctrine until 1911. At that time, technology limited American combined arms to artillery, machine-gun, and horse cavalry support for the infantry. The army had not yet integrated the very latest military technologies into published doctrine, such as motorized vehicles and aircraft, nor had American industry begun to mass produce these advanced weapons. Even after entering the Great War in 1917, factories in the United States were not ready to produce what the AEF needed to fight a modern industrialized war.

The extent to which the AEF relied upon the British and French for weapons, equipment, and vehicles attested to America's military-industrial deficiency. General Pershing placed Brigadier General Charles G. Dawes in charge of procuring these much needed weapons and supplies from the Allies. Dawes purchased approximately five thousand artillery pieces, ten thousand machine guns, four thousand mortars, forty thousand automatic rifles, six hundred tanks, and two hundred thousand horses. This chapter will explain how the AEF gradually incorporated these weapons, and many more, into a brand new combined arms tactical doctrine. Theses tactics would not only serve the AEF on the western front, but laid the foundation for more sophisticated U.S. Army combined arms during the interwar period, the Second World War, and beyond.

Artillery Doctrine

American artillery performed poorly during the Spanish-American War in 1898. The U.S. Army possessed few modern artillery pieces for training during that conflict. Therefore, most artillery crews had no experience firing the M1885 3.2-inch field guns they used in battle against the Spanish. American bombardments were consistently inaccurate. Furthermore, maneuverability issues prevented artillery crews from getting to where they were most needed to support the infantry on the battlefield. During the Punitive Expedition into Mexico in 1916, General Pershing only brought one battalion of M1902 3-inch artillery to pursue the *Villistas*. The U.S. Army never actually fired these guns in combat in Mexico, as they spent most of their time constantly on the move trying to capture General Francisco "Pancho" Villa and his *guerrillas*.

Before the Great War, most armies used artillery primarily to bombard and soften enemy positions prior to infantry assaults. Many commanders believed that even mass frontal attacks could succeed with enough artillery preparation. They quickly discovered in 1914 that even their most intense preliminary bombardments failed to reduce enemy trenches and defenses enough to allow large bodies of infantrymen to safely advance. Even if only a few enemy machine guns survived the preliminary bombardment, they were enough to decimate an entire infantry assault. Restoring offensive mobility against the modern, industrial, defensive weapons of the western front required more precise and sophisticated artillery support for the infantry. To meet this demand, the Imperial German Army introduced "standing," "box," and "drum" barrages in late 1914. A standing barrage was a sustained bombardment of an enemy position that continued until the infantry reached that position. A box barrage involved bombarding the flanks

I. AEF Combined Arms Doctrine

and rear of an enemy position to prevent retreat, reinforcement, and resupply while the infantry assaulted that position. A drum barrage was a short, rapid, preliminary bombardment designed to disorient the enemy moments before an infantry attack. German artillery support for the infantry evolved further with the "Mackensen Wedge." General August von Mackensen's Eleventh Army introduced this tactic on the eastern front in 1915. This was an intense, concentrated, heavy artillery bombardment of a specific section of the enemy line to create an opening or "wedge" for the advancing infantry to pass through.

The Imperial German Army introduced the famous creeping barrage during their attack on Verdun in February 1916. The creeping barrage involved the infantry advancing behind a protective curtain of artillery fire that preceded them all the way to their objective. Advancing the infantry behind this barrage required precise coordination and was risky, often resulting in friendly casualties. The British Army published tactical doctrine for creeping barrages in the summer of 1915 and conducted creeping barrages of their own during the Battle of the Somme. In this case, British attackers often lagged too far behind the curtain of artillery fire and German defenders had time to move back into their defensive positions after the barrage passed. The British did not perfect their creeping barrage until the Battle of Passchendaele one year later. The French Army conducted their first creeping barrages during the counterattack on Fort Douaumont at Verdun in October 1916. Neither the Germans, French, nor British would even consider launching an infantry assault without a creeping barrage by 1917. The British believed that a creeping barrage was the only way to bring the infantry within bayonet range of the enemy. The 1917 French *Manual of the Chief of Platoon of Infantry* asserted that it was impossible for infantry to capture a defensive position without a creeping barrage.

Creeping barrage coordination improved further in 1918 with the placing of artillery liaison officers in infantry battalions to communicate with supporting batteries. German artillery officer Colonel Georg Bruchmuller pioneered this technique. He went over battle plans meticulously with officers and enlisted men from both the infantry and artillery prior to every attack. Nevertheless, armies on the western front coordinated most creeping barrages with limited success using only a pocket watch and a timetable chart, advancing one hundred to three hundred meters every three to five minutes.[1]

After observing the fighting in France in the summer of 1917, Colonel Charles P. Summerall reported to the War Department on the essential role of artillery support for the infantry on the western front. General Pershing, however, harbored reservations about the battlefield progress of his infantry being limited by the artillery's ability to support it. Furthermore, he feared that friendly casualties due to poor coordination between the infantry and artillery would have a disastrous impact on American troop morale. Nevertheless, Pershing allowed for French artillery to support the AEF's attack on Cantigny in May 1918. American infantrymen used colored signal flares to direct French artillery fire during the battle. The AEF conducted an effective creeping barrage of their own during the Battle of Soissons in July 1918.

A vast amount of U.S. Army tactical literature published during the Great War discussed artillery as a combined arm. These publications described various types of artillery and their infantry-support application. AEF heavy artillery consisted of French Schneider 155-millimeter howitzers. Each division contained one heavy artillery regiment of twenty-four howitzers. Light artillery was comprised of French Schneider M1897 seventy-five-millimeter cannon. The seventy-five's high rate of fire made it ideal for conducting creeping barrages. There were two light artillery regiments of twenty-four guns in each division. The AEF also used eight-inch heavy howitzers and fourteen-inch railway-mounted naval guns for special tasks, such as bombarding railroads, tunnels, bridges, and ammunition depots anywhere from seven to twenty-four miles behind enemy lines. The U.S. Army organized its heavy howitzers into independent battalions. The AEF utilized five massive railway guns during the war.

Brigadier General Arthur L. Wagner's 1894 *Organization and Tactics* was in its eighth edition in 1917 and still widely used by the U.S. Army during the Great War. The publication clearly stated that infantry was ineffective without artillery support. It described heavy artillery reducing entrenchments and neutralizing enemy artillery. It also explained the use of light artillery in trench warfare. Wagner defined direct fire as level fire against visible targets and indirect fire as angled

I. AEF Combined Arms Doctrine

fire against targets visible only from elevated observation points. The book claimed that firing on the enemy's rear or flank was effective, but catching the enemy in crossfire from opposite flanks was best. It recommended high explosive shells with percussion fuses against fortified positions and shrapnel shells with fuses set to detonate above or in front of the enemy against enemy personnel.[2]

Infantry Drill Regulations United States Army 1911, Corrected to April 15, 1917, echoed Wagner's discussion of artillery as a combined arm. It defined neutralizing enemy artillery and defenses as the primary role of artillery during infantry assaults. The manual urged infantrymen to accustom themselves to the noise and concussion involved in advancing under the cover of artillery fire.[3]

Field Artillery Notes No. 5 of June 1917 called for artillery officers to maintain close communication with infantry officers. It claimed that the relationship between the artillery and infantry was critical to the establishment of mutual confidence in battle, a key component of combined arms. The pamphlet urged artillery officers to maintain telegraph and telephone lines to optimize this communication.[4]

Manual for Noncommissioned Officers and Privates of Field Artillery of the Army of the United States, Volume I, Corrected to December 31, 1917, explicitly identified artillery as a support arm for the infantry. The manual described various artillery shell types and firing methods for supporting the infantry. It recommended shrapnel shells against moving targets and high explosive shells for stationary objects and enemy troops inside trenches. It described "salvo fire" as short busts alternating from opposite ends of the firing line for making targeting and range adjustments. "Volley fire" represented all guns firing for a set length of time to reduce or suppress a specific enemy position. "Volley sweeping fire" involved all guns making slight adjustments between salvos to saturate large areas.[5]

Manual for Noncommissioned Officers and Privates of Field Artillery of the Army of the United States, Volume II, Corrected to December 31, 1917, also focused on artillery support for infantry attacks. It described field guns destroying defensive positions and enemy artillery prior to offensives and suppressing defenders and creating gaps in enemy lines during infantry assaults. The manual stressed the importance of silencing enemy artillery during the initial phase of an infantry attack. Fire then needed to be shifted to enemy defenders, obstacles, and barbed wire as the infantry approached enemy lines. The manual stressed the importance of communication when the infantry entered the field artillery's zone of fire.[6]

Instructions for the Offensive Combat of Small Units directed infantry officers to communicate with the artillery. The manual urged infantry company commanders to attack immediately after the artillery's preliminary bombardment and keep pace with creeping barrages.[7]

General Pershing's *Tactical Note Number 7, Combat Instructions for Troops of First Army*, encouraged the use of creeping barrages to clear enemy barbed wire for the advancing infantry. It noted how attackers needed to advance seconds behind the curtain of artillery fire to take full advantage of the barrage. The AEF recommended one hundred meters as the ideal distance between a creeping barrage and the advancing infantry. Most infantry commanders, however, utilized a more cautious three hundred-meter distance to avoid friendly fire casualties.[8]

The U.S. War Department published *Drill Regulations for the 75 French Gun Model 1897, Volume III* in September 1918, just as the AEF was about to launch the Meuse-Argonne Offensive. It defined light artillery as an infantry-support weapon for neutralizing enemy personnel and artillery. The publication proclaimed that seventy-five batteries needed to advance with the infantry to maintain effective fire support throughout an attack. Moving the guns, however, decreased accuracy and exposed them to enemy fire. This regulations recommended batteries preselect artillery positions for optimal fire and mobility. The manual prescribed moving light artillery piecemeal to enable batteries to cover each other while advancing.[9]

General Pershing never let go of his concerns about creeping barrages slowing the progress of his infantry and inflicting friendly casualties. His September 1918 *Combat Instructions* ordered close artillery cover for all infantry advances during the Meuse-Argonne Offensive, but urged better observation and communication to reduce friendly fire. Pershing's anxiety over infantry/artillery coordination was not

I. AEF Combined Arms Doctrine

unfounded. On 9 November 1918, Lieutenant Leslie Davies of the Seventy-Ninth Division's 311th Machine-Gun Battalion "saved part of his own company ... from annihilation by American artillery ... discharging flare after flare until the signaling was noted and the barrage stopped."[10]

Pershing's *Combat Instructions* also commented on heavy artillery batteries using flash ranging to provide indirect fire support for the infantry. The British had been using the sound and muzzle flash of enemy artillery to direct effective counter-battery since Passchendaele. Sound ranging observers used microphones to identify the type and location of enemy artillery pieces located up to nine hundred yards away. They telephoned this data to heavy artillery crews. Similarly, cartographers used information phoned in by flash ranging spotters to plot the location of enemy batteries on a map. Needless to say, sound and flash ranging were most effective when used in concert with aerial observation.[11]

The AEF's concept of artillery as a combined arm expressed in these manuals, regulations, and combat instructions represents significant growth from the U.S. Army's prewar view of artillery as weapon suited only to preparing the battlefield for an infantry assault. Even General Pershing moved beyond his pre–1917 accusation that the French and British army's overdependence on artillery was responsible for bogging them down in a stalemate with the Germans on the western front. In fact, Pershing was determined to increase the artillery's ability to keep pace with his advancing infantry in 1918 by outfitting all batteries with motor vehicles. Unfortunately, the AEF never realized this goal. Colonel Conrad H. Lanza, Chief of Artillery Operations for First Army during the final phase of the Meuse-Argonne Offensive, illustrated just how far the AEF was from realizing the commander-in-chief's vision for mobile artillery: "Due to the length of time required to place [artillery pieces] in firing position from traveling position they are not suited for advance guard duty ... hours would be needed to find a firing position and move the guns into them."[12]

American Tactical Advancement in World War I

Chemical Warfare Doctrine

The Germans introduced chlorine gas on the western front during the Second Battle of Ypres in April 1915. The British responded with their own poison gas attack at Loos in September 1915. Within a year, all belligerents engaged in chemical warfare. Using poison gas as an infantry support weapon was dangerous. Chemical agents lingering on the battlefield posed as much a threat to friendly troops as the enemy. The widespread use of gasmasks in 1916 and their continued improvement throughout the conflict turned chemicals into more of a battlefield nuisance than the war-winning weapon the Germans envisioned in 1915. Less than two percent of chemical agent casualties were fatalities. Most gas victims recovered and returned to combat. Tragically, the lethality of chemical weapons became more apparent after the war, when veterans developed terminal respiratory diseases from their wartime exposure to poisonous gases.

Chemical agents as a combined arm did not appear in U.S. Army tactical doctrine until late spring 1918. In 1919, however, AEF general headquarters published an extensive multivolume work covering all aspects of chemical warfare. The manuals detailed types of poisonous agents, their capabilities, methods of delivery, and coordinating chemical weapons with other arms.

Gas Manual, Part I: Tactical Employment of Gases reflected everything the U.S. Army had learned about chemical warfare during the Great War. It began by describing a multitude of delivery methods. For example, special gas troops could dispense chemicals from cylinders, Livens projectors, or four-inch Stokes mortars. The Livens projector tube delivered a sixty-five-pound explosive drum of chemicals up to eighteen hundred yards. The Stokes mortar lobbed chemical projectiles in a higher arc than the Livens projector. Artillery could fire shells filled with chemical agents over greater distances. Finally, infantrymen could deploy their own chemical-filled rifle and hand grenades. The publication then described gas tactics for supporting the infantry. For example, artillery batteries could provide screens of poisonous gas and smoke ahead of advancing infantry, deploy chemical agents around enemy positions to trap defenders during infantry attacks, and drop

I. AEF Combined Arms Doctrine

gas on enemy reserves in rear areas to prevent counter attacks. The manual recognized that chemical weapons did not always produce casualties, but they did place the enemy at a disadvantage by decreasing visibility, hampering breathing, lowering morale, and forcing the use of gas masks. Finally, the publication outlined the importance of reconnaissance and intelligence in chemical warfare, as information about weather, topography, and the enemy's anti-gas capabilities was critical when planning a chemical attack.[13]

The first volume of the gas manual categorized chemical agents and their uses as well. Phosgene was a lung irritant that produced immediate symptoms. It dissipated quickly, however, and required constant redeployment. Furthermore, gasmasks totally nullified the effects of phosgene. The publication recommended phosgene be used in combination with other agents, such as chloropicrin. Chloropicrin penetrated gasmasks, induced vomiting, and was highly persistent. The infamous mustard gas was a blister-producing gelatinous vapor/liquid that was slow to generate symptoms but remained active for up to a week. Mustard gas was ideal for making large areas uninhabitable to the enemy for extended periods of time. These areas, however, would be untenable by friendly infantry as well. Bromobenzylcyanide caused eyes to tear and was moderately persistent. Tetrachloride produced smoke. Phosphorous was an incendiary agent that smoked as it burned through enemy troops and defenses. The U.S. Army was producing what would have been the most treacherous poisonous gas of the Great War when the conflict ended. Initially developed by Captain Winford L. Lewis at Catholic University in Washington, D.C., Lewisite was an arsenic-based liquid/vapor with qualities similar to mustard gas. In addition to causing burns and blisters, however, Lewisite penetrated the skin, entered the bloodstream, and lowered blood pressure, heart rate, and body temperature. The AEF planned to use three thousand tons of Lewisite on the western front in 1919.[14]

The publication contained a section on the uses for smoke on the battlefield. It claimed that artillery was the most effective way to lay a smoke screen on the battlefield. The manual even described tanks advancing under large shrouds of smoke. Smoke bombardments like these required precision, however, as disaster could result if infantrymen or tanks became lost in their own smoke screen during an attack. The artillery could also use smoke bombardments as phony gas attacks. This would cause enemy troops to anticipate an assault in one area while the infantry advanced on another. Smoke bombs dropped from aircraft could be used to mark targets for the artillery as well.[15]

Gas Manual, Part II: Use of Gas by the Artillery specifically covered methods and uses for delivering chemical agents by artillery. "Destruction fire" was a concentrated bombardment creating a dense poisonous gas cloud, such as phosgene. The objective of destruction fire was to produce enemy casualties. "Neutralizing fire" laid a blanket of persistent chemical agents, like mustard gas. This would hinder the enemy by forcing them to wear gasmasks for an extended period of time. "Counter-battery fire" involved a burst destruction fire, followed by hours of neutralizing fire, then another burst of destruction fire. This prolonged exposure to different types of chemicals would exhaust enemy gasmasks and eventually inflict casualties. "Harassing fire" placed non-lethal agents, such as bromobenzylcyanide, behind enemy lines to disrupt troop movements. "Interdiction fire" deployed mustard gas in unoccupied terrain to prevent the enemy from moving into that area. Encircling an enemy position with mustard gas would also cut it off from resupply and reinforcements. The artillery could use mustard gas in this way to allow the infantry to bypass enemy strong points.[16]

Gas Manual, Part III: Use of Gas by Gas Troops outlined the responsibilities of AEF gas troops in support of artillery and infantry. It was possible for gas troops to release chemical agents from cylinders when positioned upwind from the enemy. The publication recommended releasing poisonous gas in this way during artillery bombardments. The gas would drive the enemy from cover and create disorder, while the artillery inflicted casualties. A short burst of phosgene from gas troops using Livens projectors, followed by an artillery bombardment, would cause the enemy to assume they were under a gas attack before being hit with shrapnel and high explosive shells. Gas troops could use Stokes mortars to provide smoke cover for infantry attacks over open ground. Gas troops could also support the infantry by using phosphorous rifle grenades to incinerate enemy machine-gun nests.[17]

Gas Manual, Part IV: Use of Gas by Infantry identified chemical

I. AEF Combined Arms Doctrine

weapons carried by infantrymen. The infantry could use phosphorous hand grenades to destroy enemy machine-gun and mortar positions or grenades bearing non-lethal agents to drive enemy soldiers out of trenches and dugouts. The manual warned against using these gas grenades when advancing into the wind. Smoke hand grenades could blind enemy machine gunners and cover small troop movements. The small amount of smoke generated by a grenade dissipated very quickly. For this reason, the publication recommended that infantrymen constantly redeployed smoke grenades until all objectives were captured. The infantry could use smoke grenades to mark targets for the artillery as well.

Despite this extensive literature pertaining to chemical agents as a combined arm, the AEF was slow to use gas against the Germans. Fear of retaliatory chemical attacks was the primary reason for this. The Americans were not desensitized to chemical agents in the way the French and British had become earlier in the war. For the less experienced U.S. Army, poisonous gas still represented a mysterious and terrifying weapon in 1918. General Pershing did issue orders for the inclusion of gas shells in all American artillery bombardments until the Saint Mihiel Offensive in September 1918.[18] This represented a drastic turning point in the AEF's usage of chemical agents. The artillery's supply of gas shells was the only limitation on its use of chemical agents during the last two months of the war. American infantry attacks were almost always supported by poisonous gas during Meuse-Argonne Offensive.

Machine-Gun Doctrine

With the exception of its brief infantry support role during the Battle of San Juan Hill, the U.S. Army only used the rapid firing Gatling gun in a defensive capacity during the Spanish-American War. The army kept over a thousand of these Gatling guns in service until 1915. In the meantime, the U.S. Army adopted a Maxim-style machine gun produced by the Colt's Manufacturing Company in 1904. The mounted cavalry convinced the army to replace this weapon with the more portable French Hotchkiss M1909 Benét-Mercié in 1911. By 1915, the army had nearly thirteen hundred Benét-Merciés. The weapon's sensitive loading and firing systems proved inadequate, however, and the U.S. Army abandoned it by 1917.[19] This left the AEF without a standard machine gun when it entered the Great War.

The U.S. Army did not fully realize the machine gun's offensive potential prior to the Great War. *Field Service Regulations 1908* made no mention of machine guns at all. The cavalry considered using machine guns in place of dismounted troopers for fire support, but the weapon's lack of mobility made this impossible. Wagner's *Organization and Tactics* categorized the machine gun as a defensive weapon due to its lack of mobility. The War Department's 1915 *Combined Infantry and Cavalry Drill Regulations for Automatic Machine Rifle, Caliber .30, Model of 1909*, only discussed the machine gun as a defensive weapon for trench warfare. The Great War would finally expand U.S. Army's machine-gun tactics to include an offensive role.

Prewar reluctance to recognize the machine gun as an offensive weapon was not unique to the U.S. Army. The French Army paid little attention to the offensive potential of the machine gun leading up to the Great War, as its commanders believed the weapon conflicted with their philosophy of *élan*. The prewar British Army was confident in the superior marksmanship of their riflemen, each firing twenty to thirty aimed shots per minute, over the inaccurate rapid fire of the machine gun. By 1917, however, French and British tactical doctrine contained guidelines for machine-gun support for the infantry. The French Army's *Manual of the Chief of Platoon of Infantry* placed machine guns at the disposal of infantry battalion commanders and occasionally assigned machine guns to platoons for special tasks. The publication suggested that the widespread use of machine guns in an attack would reduce the number of required soldiers to take the objective. According to the French manual, machine guns could be used during an assault to protect the infantry's flanks, provide a creeping barrage of machine-gun fire, and defend captured ground. The British Army increased the number of machine guns in each battalion from two in 1914 to four in 1915. The Imperial German Army assigned mobile machine guns to its assault units in 1917. This 1917 German Maxim 08/15 light machine gun weighed forty pounds and employed a stock, grip, and bipod for portability.

I. AEF Combined Arms Doctrine

The AEF finally experienced the offensive application of machine guns firsthand during the Battle of Soissons in July 1918, when machine guns supported an American infantry advance of four miles. By the end of the war, the U.S. Army had 224 machine guns assigned to each division for infantry support. This was more machine guns per division than any other army on the western front in 1918.[20]

There were two machine-gun battalions in each AEF infantry brigade and four machine-gun companies in each regiment, as opposed to one machine-gun company per regiment in the prewar U.S. Army. A machine-gun battalion contained ninety-six guns. A machine-gun company had sixteen guns. Machine-gun battalions, including one motorized machine-gun company, became a permanent part of each American division in May 1917. Most AEF divisions used the French Hotchkiss M1914 gas-operated, air-cooled machine gun. The Hotchkiss fired a strip of thirty-eight-millimeter Lebel M1886 rounds at a rate of five hundred rounds per minute. Its air-cooled design allowed for hours of continuous use. The weapon weighed 108 pounds with its tripod. Other divisions had the British Vickers water-cooled machine gun. The Vickers fired a belt of 250 .30-inch rounds at a rate of five hundred rounds per minute. The gun weighed ninety-eight pounds with its tripod. A handful of AEF divisions received the American made Browning M1917 water-cooled machine gun before the cessation of hostilities. The Browning fired a belt of .30-inch rounds at a rate of five hundred rounds per minute. The gun weighed eighty-five pounds with its tripod.

AEF tactical manuals and pamphlets outlined the coordination of machine guns with both infantry and artillery. *Provisional Machine-Gun Firing Manual, 1917* estimated that one machine gun could fire the same number of rounds per minute as sixty riflemen. A mounted machine gun's field of fire, however, was more limited than a group of riflemen. This made machine guns ideal for firing on a fixed enemy position, but not for shifting between targets. The manual warned against keeping a machine gun in the same position for an extended period of time. When a machine-gun crew opened fire, it revealed its position to enemy artillery and became an easy target.[21]

Notes on the Employment of Machine Guns asserted that a creeping barrage of indirect machine-gun fire could cover advancing infantry for three thousand yards. Machine guns were not mobile enough to be included in the first line of an infantry attack, but could be brought forward for subsequent fire support or to hold captured ground. The latter became a critical task for American machine-gun crews during the Great War, as German elastic defensive tactics called for counter attacks to be launched immediately after yielding territory. The publication recommended one machine gun be positioned every 150 yards during an attack. Each machine-gun crew could advance behind the infantry by bounds along predetermined paths. The manual also outlined various types of machine-gun fire. When providing a creeping barrage, machine guns could function like artillery by sweeping the terrain immediately ahead of advancing infantry. Machine guns had to be placed on an elevation three hundred to one thousand yards behind the infantry to provide this kind overhead fire support. "Neutralizing fire" called for machine guns to suppress fixed enemy positions during infantry attacks. The War Department pamphlet described employing machine-gun cover for artillery batteries on the move as well. During artillery bombardments of villages and towns, the manual directed machine-gun crews to focus their fire on the streets. Enemy soldiers who fled demolished structures would run into this machine-gun fire.[22]

Instructions for the Offensive Combat of Small Units claimed machine guns were ideal for suppressing enemy machine guns during infantry assaults by asserting that machine-gun crews could advance just behind the infantry throughout an attack. AEF general headquarters' *Employment of Machine Guns* described machine-gun creeping barrages utilizing both direct and indirect fire. Machine-gun crews could suppress enemy machine guns and protect the infantry's flanks during simultaneously. Commanders had to remember to constantly order machine guns forward to maintain this continuous cover for the infantry. When available, machine-gun crews could use mules and carts to speed their advance.[23] In theory, every divisional machine gun had its own cart and mule and independent machine-gun battalions had their own motor transport. Mule and automotive shortages made this impossible during the Meuse-Argonne Offensive.

General Pershing's *Tactical Note Number 7* depicted the machine

I. AEF Combined Arms Doctrine

gun as an infantry support weapon. He urged his battalion commanders to direct their machine-gun companies to provide covering fire and suppress enemy strong points during infantry assaults. Pershing's *Combat Instructions* ordered two machine guns to follow each attacking platoons by advancing in bounds with one gun providing cover for the other. The commander-in-chief intended to increase the mobility of machine guns by assigning superlight three-ton tanks to each machine-gun battalion. These vehicles would pull machine guns across the battlefield and support them with firepower.[24] The armistice prevented this from materializing.

Mortar Doctrine

The mortar was not a new weapon in the U.S. Army in 1917, but the Great War turned it into a common tool on the battlefield. Mortars played a prominent role in French and British trench-warfare tactics by 1916. The mortar's high arching fire was ideal for lobbing shells into enemy trenches from covered positions. Its portability allowed the infantry to carry its own "trench artillery" support as it advanced. For example, German attackers pulled light mortars on wheels during the General Erich Ludendorff's Offensive of spring 1918. U.S. Army tactical doctrine envisioned a similar infantry support role for mortars during the Great War. Each AEF division had one battery of thirty-six mortars, with twelve mortars in a brigade and six in a regiment. There were no mortars on the company level. The U.S. Army used British three-inch Stokes mortars and six-inch Newton mortars during the war. The Stokes lobbed an eleven-pound projectile up to one thousand yards. The Newton fired a twelve-pound projectile up to two thousand yards. The AEF attached its Newton mortar crews to artillery brigades.

Instructions for the Offensive Combat of Small Units noted that the mortar's mobility allowed it to provide infantry support throughout an attack. The manual described mortar crews directing high angle fire into enemy trenches and foxholes during infantry assaults. In *Tactical Note Number 7*, General Pershing urged battalion commanders to direct their Stokes mortars at enemy strong points during attacks. His *Combat Instructions* recommended Stokes-mortar crews work in pairs to cover each other while advancing with the infantry. When trucks were available, they could move the heavier Newton mortars forward with the infantry as well.[25]

Grenade Doctrine

Handheld bombs first appeared in Europe in the fifteenth century. Modern percussion grenades debuted during the Russo-Japanese War and the Balkan Wars of 1912 and 1913. For the British and French during the Great War, the grenade proved itself early on as an ideal weapon for trench raiding, dislodging machine guns, and harassing entrenched enemy troops. Some of the first grenades on the western front were ration cans packed with explosives and nails and rigged with makeshift fuses. Eventually, a wide variety of industrially produced grenades appeared, including ball, egg, and stick types. The British made use of over ten different kinds of grenades by 1915 and required all of its soldiers to have grenade training. Despite this, British commanders preferred to group grenadiers into specialized units instead of arming every infantryman with his own grenades. The British made wide use of these grenadier units during the Battle of the Somme. By 1916, the French Army administered basic grenade instruction to all its troops. Specialized grenadiers, however, underwent more sophisticated training in a variety of grenades, tactics, and the rifle grenade. The French *Manual of the Chief of Platoon of Infantry* recommended raiding parties deploy grenades prior to entering an enemy trench. Grenades could be used while the raiding party maneuvered through the enemy trench as well. The manual also described rifle grenadiers creating a creeping barrage of grenades for advancing infantry.[26]

Similar to mortars, grenades were not new in the U.S. Army in 1917, but they found a standard role in the army during the Great War. The U.S. Army viewed the grenade as a weapon for driving enemy troops out of cover and into the open. Like the British and French, the AEF trained every soldier in grenade use, but did not issue grenades universally. Instead, each American infantry company contained four sections of twelve grenadiers and four sections of nine rifle grenadiers. Commanders dispersed grenadiers throughout the company as they

I. AEF Combined Arms Doctrine

saw fit. Company commanders usually reserved their grenadiers for specialized task, such as trench raids. The AEF utilized the American made Mark III high explosive concussion grenade. The British Mills bomb, French LeBlanc, and American Mark II served as the AEF's fragmentation grenades. Most of these grenades possessed a four to seven-second fuse and a blast radius of about eighty feet. The concussion grenade was the preferred offensive grenade, as shrapnel from fragmentation grenades posed a threat to advancing friendly troops.

Notes on Grenade Warfare: Compiled from Data Available on February 15, 1917, Army War College defined the grenade's role in trench raiding and provided tactics for such raids. Trench-raiding patrols were ideal for gathering intelligence just prior to an attack. Raiding parties took prisoners for interrogation and collected information regarding enemy defensive capabilities. According to the publication, the grenade, not the rifle, was the primary trench-raiding weapon. A grenade could be thrown while running or lobbed over the heads of friendly infantry. It also had a wider range of destruction than a rifle. Trench-raid tactics were a microcosm of combined arms. Based on the French Army's model, eight men composed an AEF raiding party, this included riflemen, grenade carriers, and grenade throwers. Each party had 120 grenades. The riflemen's role was to protect the party's flanks and spot for the throwers. A carrier with a bag of grenades accompanied each thrower during the raid.[27]

Artillery support for trench raids required accurate knowledge of terrain and precise timing, as creeping barrages needed to cease when raiding parties entered the enemy trenches. Raiders could use flares to signal their arrival to artillery crews. The AEF made use of Very flare pistols for this type of signaling. The publication recommended throwers toss numerous grenades into enemy trenches before entering. Grenadiers could use teargas or phosphorous on enclosed enemy positions, such as dugouts and bunkers. When appropriate, grenadiers could use mechanical throwers to provide preliminary bombardments and creeping barrages for trench raids. Mechanical throwing devices relied on springs or compressed air to launch grenades up to 350 yards. Grenadiers could set the machines to repeatedly deploy grenades on a designated position.[28]

Instructions for the Training of Platoons for Offensive Action, 1917 outlined tactics for the use of grenades during infantry assaults. The manual rated the grenade as the infantryman's third most important weapon, behind the rifle and bayonet. It described rifle grenadiers firing grenades up to two hundred yards from behind cover to drive enemy soldiers into the open during infantry assaults. The French introduced rifle grenade launchers on the western front in 1915. American rifle grenadiers used French Vivien-Bessière rifle grenades and cup launchers. This system used the gas discharged from a standard rifle round to arm the grenade and propel it from atop the rifle muzzle. Rifle grenadiers fired their rifles at a forty-five-degree angle with the stock planted firmly on the ground. *Instructions for the Offensive Combat of Small Units* described the use of grenades and rifle grenades by platoon leaders in the first and second assault lines to reduce points of enemy resistance. General Pershing's *Combat Instructions* recommended grenades be use in fire-and-maneuver tactics to destroy suppressed enemy machine guns.[29]

Automatic Rifle Doctrine

The French Army debuted the automatic rifle in 1915 for additional firepower. The French and British initially deployed automatic rifles like machine guns, as primarily defensive weapons. By late 1915, however, the French and British came to appreciate the automatic rifle's portability and added the weapon to their offensive units for fire support. The French Army's *Manual of the Chief of Platoon of Infantry* included automatic rifles in the first infantry assault wave to neutralize enemy defensive positions for subsequent attack waves. The automatic rifle became a standard infantry support weapon in every British platoon by 1916. With its portability and a rate of fire between a rifle and a machine gun, the weapon found a similar place in the AEF. Each American infantry company had four sections of fifteen automatic riflemen. A single soldier operated the rifle. Although the AEF assigned two ammunition carriers to each automatic rifle, the weapon did not require support personnel. This set the automatic rifle apart from the machine gun, which required at least two operators. Like grenadiers,

I. AEF Combined Arms Doctrine

company commanders attached automatic riflemen to platoons for special tasks, such as neutralizing enemy machine guns.[30] By the end of the Great War, the automatic rifle earned its place as the U.S. Army's standard, small-unit, fire-support weapon.

Instructions for the Training of Platoons for Offensive Action described automatic riflemen providing mobile covering fire for infantry units conducting fire-and-maneuver tactics. Automatic riflemen could fire from cover and suppress enemy troops while infantrymen flanked with grenades and rifles. *Notes on the Employment of Machine Guns* identified situations where commanders preferred using automatic rifles over machine guns, as automatic rifles were better suited for keeping pace with advancing infantry and providing more direct fire support.[31]

The War Department translated and published the French *Manual of the Automatic Rifle (Chauchat), Drill-Combat-Mechanism* in April 1918. The French M1915 Chauchat weighed twenty-two pounds, had a curved magazine with a twenty-round capacity, and could discharge 250 rounds per minute. This weapon was the AEF's most commonly used automatic rifle. The publication outlined the Chauchat's potential as an infantry support weapon when conducting fire-and-maneuver tactics. For example, it could be fired either directly on enemy positions while the infantry approached from the flanks, or indirectly over the heads of advancing infantry. Automatic riflemen could also draw enemy fire away from attacking infantrymen. Automatic rifles were mobile enough to be included in flanking attacks as wells.[32] The translated French manual described the preferred firing position for an automatic rifle as prone; however, automatic riflemen could also fire while advancing. Automatic riflemen advancing in an assault could maintain a heavy volume of fire on enemy troops in the open.[33]

Instructions for the Offensive Combat of Small Units considered automatic rifles ideal for mobile suppressing fire at short and medium ranges. The instructions described assaulting infantry platoons directing automatic rifle fire against multiple enemy strong points during the course of an attack. General Pershing's *Combat Instructions* detailed the combined use of automatic rifles and grenades for eliminating enemy machine-gun positions during offensives.[34]

American Tactical Advancement in World War I

Small Arms and Special Weapons Doctrine

Despite the fact that American soldiers made little use of bayonets during the Spanish-American War, the weapon continued to hold a prominent place in U.S. Army tactics in the early twentieth century. For example, General Pershing still viewed the rifle and bayonet as the infantryman's most important tools on the battlefield during the Great War. This line of thinking was not unique to Pershing. Successful Japanese *banzai* charges during the Russo-Japanese War convinced most European armies that the bayonet still had a role in modern warfare. The bayonet appeared in French and British tactical doctrine and adorned the rifles of soldiers of all armies throughout the Great War. French and British tactics recommended fixed bayonets whenever the infantry advanced. The French believed that the bayonet, more than any other weapon, gave attacking soldiers the courage to keep moving forward, and British tactics relied on the bayonet when closing with the enemy.

Nearly half the AEF used bolt-action, magazine-fed M1903 Springfield rifles with sixteen-inch M1905 bayonets. The rest carried British, bolt-action, magazine-fed M1917 Enfield rifles with seventeen-inch M1917 bayonets. *Instructions for the Training of Platoons for Offensive Action* reflected Pershing's strong sentiment about the rifle and bayonet. According to the manual, the purpose of combined arms was to aid the infantry in getting close enough to engage the enemy with the bayonet. *Manual for Noncommissioned Officers and Privates of Infantry of the Army of the United States, 1917* stated, "How to use the bayonet, and the will to use it, must often be the deciding factors in battle."[35]

Small Arms Firing Manual 1913, Corrected to March 15, 1918, outlined the War Department's terminology and expectations for marksmanship. "Fire effectiveness" was the number of targets hit over a set length of time. "Fire superiority" meant maximizing the number of targets hit within a specific time. Attaining fire superiority was challenging, however, since increasing the rate of fire decreased accuracy. The War Department expected American soldiers to be able to fire ten aimed shots per minute at a range between two and four hundred yards,

I. AEF Combined Arms Doctrine

seven shots between five and seven hundred yards, and five shots between eight hundred and one thousand yards. In comparison, the French Army's *Manual of the Chief of Platoon of Infantry* limited riflemen to an eight-shot per minute rate of fire and a maximum range of only 650 yards. On the western front, the French Army reduced this to a mere fifty-five yards.[36] The discrepancies in American and French marksmanship standards reflected the General Pershing's confidence in the infantryman and his rifle, as well as the U.S. Army's lack of experience with the deadly realities of modern industrialized warfare.

Instructions for the Offensive Combat of Small Units maintained Pershing's assertion about primacy of the rifle and bayonet, but added special weapons to the formula. The publication included the French M1916 thirty-seven-millimeter cannon, or one pounder, for direct fire support for attacking infantry. The small cannon required a crew of three, fired at a rate of twenty rounds per minute, and had a range of twenty-five hundred yards. The gun employed armor-piercing projectiles against machine-gun positions and shrapnel canisters against enemy troops. Each AEF regiment contained three, thirty-seven-millimeter cannons. Commanders usually placed them in the second assault line. The three-man crews usually carried the weapon, as mules were typically unavailable for this task. Crews needed to position, fire, and move the cannon rapidly to avoid detection by enemy artillery. *Instructions for the Offensive Combat of Small Units* also mentioned the use of flamethrowers as an infantry support weapon. The publication recommended flamethrowers for clearing enemy trenches and dugouts and securing the infantry's flanks by setting fire to surrounding terrain.[37]

The Imperial German Army introduced flamethrowers on the western front in February 1915. While the German and French armies utilized specially trained flame troops, the British issued flamethrowers to ordinary infantrymen. The AEF employed this practice as well. For this reason, American flamethrower operators tended to be a danger to friendly troops. These inexperienced laymen carried combustible fuels and pressurized propellants into the combat zone. Needless to say, the U.S. Army seldom employed flamethrowers during the Great War.

American Tactical Advancement in World War I

General Pershing's wartime notes and instructions eventually encouraged American infantrymen to use special weapons alongside their rifles and bayonets. His *Combat Instructions* recommended thirty-seven-millimeter cannon crews advance in pairs during infantry assaults, with one gun suppressing the enemy while the other moved forward.[38] During the last day of the Meuse-Argonne Offensive, infantry columns of the 65th Brigade of the 33rd Division used thirty-seven-millimeter cannons in this way to support their advance against German machine-gun positions around the town of Pintheville.[39]

Tank Doctrine

The mechanical failure of the first British tank assault on the Somme did little to impress the U.S. Army with the importance of armored fighting vehicles in modern warfare. Aside from a few prototypes, the army possessed no tanks in 1917. In November of that year, 324 tanks supported the British breakthrough at Cambrai. This success convinced General Pershing of the potential of mass tank attacks. The commander-in-chief envisioned one thousand tanks supporting his infantry divisions on the western front. Needless to say, American industry never met this expectation during the Great. French tanks supported American operations at Cantigny and Soissons. Lieutenant Colonel George S. Patton Jr.'s First Tank Brigade participated in the Saint Mihiel Offensive on 12 September. Patton's brigade of 144 light tanks supported First Army's IV Corps' twelve-mile advance during the first two days of the battle, and clashed with Imperial German troops on the third. Most of the American tanks became stuck in the mud, broke down, or ran out of fuel before the battle ended on 15 September.

The U.S. Tank Corps comprised of two battalions of French Renault FT17 light tanks. There were four companies per battalion with four platoons in each company. A full-strength platoon consisted of five tanks. Operated by a two-man crew, the Renault weighed 14,500 pounds and had sixteen-millimeter armor. Its thirty-five horsepower gasoline engine produced a speed of five miles per hour. Renaults were armed with either a Hotchkiss machine gun or a thirty-seven-millimeter cannon. The American tank corps also included one battalion of British

I. AEF Combined Arms Doctrine

Mark IV and Mark V heavy tanks. Operated by an eight-man crew, these tanks weighed thirty tons apiece and had twelve-millimeter armor. They traveled four miles per hour with 105 horsepower gasoline engines. Their armament consisted of various combinations of Hotchkiss machine guns and thirty-seven-millimeter cannons. This heavy tank battalion served with the American divisions under British command.

In theory, General Pershing's vision of mass armored assaults called for heavy tanks to create paths through the enemy defenses while light tanks penetrated the gaps. The AEF never possessed enough tanks to carry out an operation of this scale, but tanks did appear in an infantry support role in published U.S. Army tactical doctrine during the Great War.[40] *Organization and Tactics* treated tanks as mobile artillery. According to the manual, tanks provided the solution to the artillery's inability to keep pace with advancing infantry. *Instructions for the Offensive Combat of Small Units* recommended tank support for infantry assaults whenever topography permitted. Tanks could create openings for the infantry with their mobile firepower and ability to flatten enemy barbed wire. The publication warned against tanks advancing beyond the infantry or venturing onto the battlefield alone. The infantry had to protect tanks against enemy machine guns, anti-tank guns, and grenadiers. A bundle of grenades under a tank's hull could easily destroy its treads. Furthermore, it was necessary for the infantry to secure the tanks' territorial gains. Pershing referred to mutual support between tank crews and infantry in his *Tactical Note Number 7*. The commander-in-chief instructed infantrymen to advance no less than one hundred yards behind tanks and help tank crews direct their fire against the enemy with signal flags.[41] Armor support for American infantry during the Meuse-Argonne Offensive was inconsistent due to the limited number of serviceable tanks, but armor had found a place in the U.S. Army.

Air Support Doctrine

Like other armies, the U.S. Army debated the role of aircraft in warfare prior to the Great War. The Aviation Section of the U.S. Army

American Tactical Advancement in World War I

Signal Corps, soon to be the U.S. Army Air Service, had only twenty-one aircraft in 1915. The British, French, and Germans armies each had between 250 and 500 aircraft on the western front that same year. General Pershing brought the First Aero Squadron into Mexico with him during the Punitive Expedition in 1916. This squadron of eight Curtiss JN3 and N8 aircraft flew over five hundred reconnaissance sorties searching for "Pancho" Villa over eleven months against unpredictable desert winds and over steep mountainous elevations. Keeping with the National Defense Act of 1916's projections, the War Department expected to attach one aero squadron to each infantry division when the United States entered the Great War. The air service contained only fifty-five aircraft and America's aircraft industry was in its infancy at that time, so the War Department abandoned this ambitious plan. From then on, division commanders worked with independent aero squadrons to coordinate tactics on the western front. As a result, infantry support became an integral role of American airpower during the Great War.[42]

As with most other advanced weapons, the AEF relied on the Allies for its aircraft. Most American pilots flew French Spad VII, Spad XIII, or Nieuport 28 single-seat biplanes. A lesser number of American pilots flew British Sopwith Camel F1 single-seat biplanes. For ground support operations, the air service used French Breguet 14, British DeHavilland-4, and DeHavilland-9 two-seat biplanes for light bombing. The Breguet 14 carried a 660-pound bomb load. The DeHavillands each carried a 460-pound bomb load. The Italian Caproni CA5 was the air service's heavy bomber. Operated by a four-man crew, the massive Caproni CA5 biplane carried a 1,170-pound bomb load. French Salmson 2A2 two-seat biplanes and DeHavilland-4s served as American reconnaissance aircraft. The Salmson's camera was housed in its fuselage. The DeHavilland's camera was clamped to an exterior rack. The French-made 520-millimeter aerial observation camera was capable of taking photographs from fifteen thousand feet above the battlefield.

Prewar U.S. Army air doctrine focused exclusively on aerial reconnaissance as the way in which aircraft could support ground forces. Like its French, British, and German counterparts, the American air

I. AEF Combined Arms Doctrine

service never published tactical doctrine dealing with fire support for ground forces during the Great War. Nevertheless, American pilots flew thousands of missions in support of infantry divisions in combat on the western front. Disputes existed within the air service about how to do this effectively. Most squadron commanders simply told their aviators to strafe and bomb enemy targets of opportunity on the battlefield whenever possible. Lieutenant Colonel William L. Mitchell served as Pershing's Chief of AEF Air Service from June until August 1917 and Chief of First Army Air Service later in the war. He was not a proponent of this kind of close air support for ground forces. He believed bombing assets behind enemy lines would make a greater contribution to AEF victory by crippling Imperial Germany's industrial capacity to wage war. This kind of thinking evolved into his famous theories on long-range strategic bombing in the 1920s.[43]

The U.S. Army Air Service did not possess enough aircraft to function as an offensive arm, let alone meet Mitchell's ambitious bombing expectations, until fall 1918. The Saint Mihiel Offensive represented the largest show of American airpower of the war with approximately fifteen hundred aircraft and thirty thousand air service personnel, including ten observation squadrons, one night observation squadron, three bomber squadrons, and fourteen airfields. The U.S. Army Air Service finally served as a true combined arms component during this operation. American air forces gained air superiority and provided thorough reconnaissance prior to the attack. The Air Service supported the ground assault on the Saint Mihiel salient by bombing German aerodromes, strafing enemy troop concentrations, and even providing Patton's tank brigade with close air support. Bombers dropped seventy-five tons of ordnance on the Imperial German Army in four days.

Engineer Doctrine

There were only three engineer battalions in the entire U.S. Army in 1915. Pershing's Punitive Expedition into Mexico contained two engineer companies to build and repair his four hundred-mile supply roads. The War Department expanded engineer regiments from one to two battalions in May 1917, and each AEF divisions had its own attached engineer regiment during the Great War. Pershing suggested each infantry company be assigned a team of engineers in *Tactical Note Number 7*. The publication described engineers supporting both infantry and tank companies across the battlefield.[44] Engineer teams could use wire cutters and explosives to clear enemy barbed wire during infantry and tank assaults. American engineers often used the British Bangalore torpedo for demolishing barbed wire. A Bangalore torpedo consisted of explosives affixed to five feet of threaded pipe. When necessary, engineers could attach additional pipes to increase the length of the torpedo. By sliding the pipes along the ground, engineers could insert the explosive end into a barbed wire belt and detonated it. The blast was capable of creating a gap five feet wide in the wire.

The AEF's progress during the Meuse-Argonne Offensive relied upon engineers to remove obstacles, lay railroad tracks, construct bridges, and build and repair roads. The 304th Engineer Regiment built nine footbridges and one twenty-six-foot trestle across the Thinte River near Gibercy during the night of 10 November so that the 158th Brigade of the Seventy-Ninth Division could cross the river the next day. The engineers completed this task in less than eight hours using only ropes and lashings to mask the sound of their work from German machine gunners on the opposite bank of the river. Fifteen members of this engineer regiment won division citations for this achievement.[45]

Mounted Cavalry Doctrine

The future of mounted cavalry in modern industrialized warfare was in doubt by the late 1800s. The prominent role of dismounted British cavalry during the Second Boer War from 1899 to 1902, however, convinced the U.S. Army to retain its horse cavalry arm in the early twentieth-century. Many American tacticians envisioned horses merely carrying cavalrymen to areas on the battlefield where they could dismount and reinforce the infantry. In this capacity, cavalrymen would no longer fight on horseback or require sabers. The army had fifteen cavalry regiments with a total of about fifteen thousand officers and enlisted men in 1915. Six of these regiments accompanied Pershing into Mexico in 1916. Mounted cavalry proved ideal for pursuing *Villistas*

I. AEF Combined Arms Doctrine

across the deserts of Northern Mexico. For example, elements of the Seventh Cavalry clashed with over five hundred Mexican *guerrillas* at San Geronimo, Mexico. Nevertheless, the War Department looked upon horse cavalry as nothing more than a means of transportation by 1917.

Wartime revisions in French tactical doctrine recommended cavalry units be dismounted and used to reinforce the infantry divisions. The French Army retained only a small number of mounted cavalry units to run reconnaissance and guard the infantry's flanks. The British reduced their mounted cavalry from 130,000 troopers in 1916 to 85,000 in 1918. The Allies dissuaded the War Department from forming entire cavalry divisions in 1917. Upon entering the Great War, the War Department posted a few cavalry regiments along the Mexican border and folded the rest into artillery or machine-gun units, where their horses became pack animals. General Pershing, himself a former cavalry officer, insisted that mounted cavalry be included in the AEF. As a result, four incomplete American cavalry regiments went to France. Only one of these partial regiments actually participated in combat. Ironically, the U.S. Army experienced a horse shortage in 1918 due to the massive number of horses the United States provided the French and British prior to entering the war. Needless to say, the army paid little attention to updating cavalry tactics to suit modern industrialized warfare. *Instructions for the Offensive Combat of Small Units* asserted that mounted cavalry was ideal for pursuing enemy troops only after the infantry expelled them from their defenses.[46] The Great War only confirmed that the days of horse cavalry in the U.S. Army were numbered.

Conclusion

By the final phase of the Meuse-Argonne Offensive in November 1918, AEF tactical doctrine prescribed the combined use of infantry, field artillery, chemical agents, machine guns, mortars, grenades, automatic rifles, small arms, and special weapons. American tactical pamphlets and manuals discussed the integration of more sophisticated arms as well, such as tanks, aircraft, engineers, and even horse cavalry.

American Tactical Advancement in World War I

These publications described field artillery supporting the infantry with preliminary bombardments and creeping barrages. Chemical weapons manuals discussed chemical agents in support of infantry and other arms. U.S. Army doctrine realized the machine gun's ability to provide cover fire for advancing infantry. AEF tactical pamphlets highlighted a variety of new uses for mortars and grenades as infantry support weapons. Tactical publications recommended that the infantry's rifles and bayonets be supplemented with special weapons, such as automatic rifles and thirty-seven-millimeter cannons. American doctrine integrating tanks and aircraft began to appear in late 1918 as well. Many of the weapons outlined in this chapter did not even exist in the pre–1917 U.S. Army. This made the AEF's ability to bring these arms into its tactical doctrine by November 1918 all the more remarkable. In less than two years, the AEF paved the way for the next one hundred years of U.S. Army combined arms development.

11

AEF Open Warfare Doctrine

By the last phase of the Meuse-Argonne Offensive in November 1918, AEF tactical doctrine endorsed General Pershing's open warfare concept over the deadly mass frontal attacks of the previous century. For example, the *Manual for Noncommissioned Officers and Privates of Infantry of the Army of the United States, 1917* emphasized brisk, erratic, perpetual forward movement, and the use of cover.[1] Assault columns supplanted linear attacks. American tactical literature called for fire-and-maneuver when traversing the battlefield. Publications recommended bypassing and flanking enemy strong points. The U.S. Army encouraged commanders to take the initiative, improvise, and focus on limited strategic objectives. By the war's end, AEF doctrine even brought open warfare and combined arms tactics together. The U.S. Army's adoption and expansion of Pershing's open warfare doctrine continued into the interwar years and the Second World War. Its essence is still present in American tactics in the twenty-first century.

The Genesis of Pershing's Open Warfare Doctrine

The impact of industrialization on Civil War battlefields dominated American military thought in the late 1800s. New industrial technology granted tremendous advantages to the defender. The U.S. Army grappled with the tactical challenge of crossing "no man's land" or "the deadly ground" well before 1917. Lieutenant Colonel Emory Upton critiqued the costly mass linear assaults of the American Civil War in his 1867 manual, *A New System of Infantry Tactics*. Upton prescribed dividing long lines of infantry into smaller units to increase maneuverability, tactical versatility, and survivability on the modern industrialized battlefield. American tacticians, like General Arthur Wagner, added flank attacks and advancing in rushes to Upton's smaller flexible formations in the 1880s. The U.S. Army incorporated these concepts into its first service-wide tactical doctrine in 1891 based a manual published in Fort Leavenworth, Kansas. The infantry manual instructed commanders to replace mass battle lines with small dispersed groups of skirmishers. The publication instructed attackers to use cover and advance in rushes. These tactics saw limited use during the latter part of the Indian Wars.

Wagner's *Organization and Tactics* placed the evolution of open warfare in a historical context in 1894. According to the manual, improvements in the accuracy and rate of rifle fire since the American Civil War rendered linear attack formations obsolete. Wagner asserted that frontal assaults could only be conducted in modern industrialized warfare when enemy defenders were significantly outnumbered. He, therefore, recommended assault columns and skirmish lines. The U.S. Army had the opportunity to test these new tactics during the Spanish-American War. American troops employed irregular formations, fire-and-maneuver, and flank attacks during the Cuban and Puerto Rican campaigns in 1898. This did not, however, represent service-wide exposure to the new tactics. The Spanish-American War was too short, and only about fifty-two thousand American soldiers actually experienced combat during that conflict.

The reforms of U.S. Secretary of War Elihu Root from 1899 until 1904 provided the U.S. Army with a new framework for standardizing and disseminating service-wide doctrine by forming the Command and General Staff School at Fort Leavenworth and the War College in Washington, D.C. Furthermore, the "Root Reforms" held the National Guard to the same instructional standards as the Regular Army. The

II. AEF Open Warfare Doctrine

Great War would be the first time the U.S. Army had a general staff in charge of all army doctrine and training. General Pershing's attempt to establish an open warfare concept for the entire AEF was part of this new trend. In 1917, French and British doctrine remained grounded in trench-warfare tactics, focusing on methods for advancing from one trench to the next in the face of intense defensive firepower. Pershing associated these trench-warfare techniques with stalemate. He sought to understand how trench warfare developed on the western front and to reverse it with open warfare. To the commander-in-chief, open warfare was synonymous with victory through continuous offensive movement.[2]

Flexible Formations

French frontal assaults fared badly against German, breech-loading, bolt-action rifles during the Franco-Prussian War in 1870 and 1871. As a result, the French added flexible formations, such as skirmish lines, to their service regulations in 1875. The French Army reintroduced massive linear frontal assaults into their tactical doctrine by 1900. French commanders at the time believed that mass formations bolstered their army's offensive spirit. They also felt that French conscripts lacked the training and discipline needed to operate in small units. Massive linear formations allowed French commanders to exercise better control over their huge army of undertrained and inexperienced citizen soldiers in 1914. The French finally reintroduced columns and skirmish lines for the purposes of maneuverability and survivability in 1917 with publications like the *Manual of the Chief of Platoon of Infantry*.[3]

The British Army began incorporating flexible formations into their tactical doctrine after the Boer Wars. Successful Japanese frontal assaults during the Russo-Japanese War, however, convinced British commanders that massive linear formations remained a viable option on the modern industrialized battlefield. The British, therefore, retained mass linear formations in their tactical repertoire as they entered the Great War in 1914. Lieutenant General Ivor Maxse was a major proponent of irregular formations in the British Army early in the war. He advocated for platoons to advance in column, square, or diamond shaped formations with five to six yards between each soldier.

He recommended using fire-and-maneuver to rush from one position of cover to the next. Maxse made flexible formations a permanent feature of British doctrine and training when he became Inspector General of Training in June 1918.[4]

The Indian Wars, Spanish-American War, and the Philippine Insurrection were either too short or too unconventional to allow the U.S. Army to perfect flexible attack formations. Carrying on the army's Uptonian reformation fell to Pershing and the AEF in 1917. U.S. Army tactical publications expanded upon Upton's ideas during the Great War to offer alternatives to deadly frontal assaults. These manuals and pamphlets discouraged mass linear formations and endorsed attack columns and small flexible configurations. *Infantry Drill Regulations United States Army 1911* warned against charging in dense masses and discouraged frontal assaults. The publication suggested platoons breakup into rifle sections, advance in columns, and then disperse into even smaller skirmish lines when within 250 yards of the enemy.[5]

Instructions for the Offensive Combat of Small Units prescribed small flexible columns advancing in single or double file when traversing obstructed ground under enemy artillery fire. General Pershing's *Tactical Note Number 7* note discouraged frontal assaults. It claimed that columns were easier to control and suffered fewer casualties than linear formations. The commander-in-chief promoted irregular infantry formations again when he outlined his open warfare concept in his September 1918 *Combat Instructions*.[6] Pershing's effort to replace mass linear assaults with attack columns and flexible formations was an extension of Upton's revolutionary tactics for survival on the modern industrialized battlefield from fifty-one years earlier. AEF units, like the Sixty-Fifth Brigade of the Thirty-Third Division, formed columns to attack and capture German strong points embedded in small towns and villages across the French countryside, like Pintheville, during the Meuse-Argonne Offensive.[7]

Fire-and-Maneuver Doctrine

Turn of the century American conflicts, such as the Indian Wars, Spanish-American War, Philippine Insurrection, and the Punitive

II. AEF Open Warfare Doctrine

Expedition into Mexico, were wars of maneuver. General Pershing's experience in these conflicts led him to believe that maneuver was the key to breaking the trench-warfare stalemate on the western front in 1917. Modern industrial weaponry, however, placed the advantage with the defender during the Great War. Therefore, Pershing eventually came to appreciate the necessity to support maneuver with firepower. Fire-and-maneuver involves one group of soldiers directing firepower against an enemy position while another group maneuvers to converge upon that position from the flank or rear. The fire element suppresses the enemy as the maneuver element attacks and neutralizes the enemy.

The French and German armies used groups of skirmishers to lay down covering fire while others closed on the objective during the Franco-Prussian War. The British employed short attack rushes with suppressing fire during the Boer Wars. Early in the Great War, German assault teams took turns advancing. One team drew fire while the other moved forward. By 1917, staying low to the ground and advancing in short intervals was a standard component of British and French tactical doctrine. In *Manual of the Chief of Platoon of Infantry*, the French Army called for attackers to support one another with suppressing fire as they moved from cover to cover until they were close enough to engage the enemy with the bayonet. The French, therefore, organized its platoons into groups of seven to thirteen men for fire-and-maneuver. They also incorporated automatic rifles and grenades into their tactics for suppressing fortified defenses, such as machine-gun pits and bunkers. One group of automatic riflemen and one group of grenadiers acted as the fire component. Two groups of riflemen served as the maneuver element.[8]

Based on the British and French models, American platoons were designed for fire-and-maneuver during the Great War. The platoon was the AEF's smallest unit of infantry, as squads were only *ad hoc* infantry formations in most armies at the time. For example, French commanders occasionally divided platoons in half to create smaller fighting units as needed. At full strength, an AEF platoon mustered fifty-nine soldiers. Manpower shortages forced Pershing to reduce his platoons to forty-three soldiers in September 1918. An American platoon contained sixty-eight riflemen. Sections of seventeen riflemen served as the platoon's maneuver element. Sections of automatic riflemen and grenadiers made up the platoon's fire element. Platoon commanders deployed these sections according to the demands of a particular combat situation, such as pairing riflemen with grenadiers to form a raiding party.

Published AEF platoon-level infantry tactics included the use of fire-and-maneuver. Wagner's *Organization and Tactics* described fire-and-maneuver, the use of cover, and "shock." "Shock" meant closing with the bayonet and engaging the enemy in hand-to-hand combat. According to the manual, infantrymen could advance twenty yards per minute using fire-and-maneuver. If necessary, attackers could find natural cover on the battlefield or create it with entrenching tools. Taking cover, however, did not mean ending the assault. *Infantry Drill Regulations United States Army 1911* urged platoons to always move forward from cover to cover, even if it meant crawling.⁹

Manual for Noncommissioned Officers and Privates of Infantry of the Army of the United States, 1917 claimed, "The Best way to hold down the fire of the enemy ... is to bring the position he occupies under well conducted and continued fire." Ironically, the U.S. Army was slow to accept the notion of soldiers firing at will. Commanders in the nineteenth century believed that allowing soldiers to fire freely would lead to massive wastages of ammunition. American troops were still firing in volleys during the Spanish-American War. *Infantry Drill Regulations United States Army 1911* was one of the first American publications to endorse firing at will instead of volley fire. It defined "fire discipline" as increasing the rate of fire to attain fire superiority and reducing the rate of fire when it became necessary to conserve ammunition. Riflemen achieved "fire superiority" by firing rapidly while maintaining accuracy. Covering fire had to be heavy enough to suppress the enemy, but not indiscriminate or prolonged to the point of endangering the assault rush. In comparison, the French Army continued to dissuade soldiers from firing independently during the Great War. The French *Manual of the Chief of Platoon of Infantry* equated "fire discipline" with a commanding officer's ability to direct effective volley fire.¹⁰

Both *Infantry Drill Regulations United States Army 1911* and *Manual for Noncommissioned Officers and Privates of Infantry of the Army*

II. AEF Open Warfare Doctrine

of the United States, 1917 identified maintaining discipline under fire and finding cover as essential elements for a successful infantry advance. Within eight hundred yards of the enemy, riflemen needed to provide covering fire for flanking maneuvers. The number of riflemen providing covering fire had to be larger than the number of soldiers in the flanking assault rush. At the same time, commanders needed to insure that there were enough soldiers in the assault element to secure the objective. Rushes could vary from thirty to eighty yards depending upon the distance from cover to the objective. The publications recommended bayonet charges for the final four hundred yards of an infantry assault.[11]

Instructions for the Training of Platoons for Offensive Action directed platoons to achieve fire superiority, use flanking maneuvers, and then close with the bayonet. The War Department's *Manual for Noncommissioned Officers and Privates of Infantry of the Army of the United States, 1917* described attackers moving forward by rushing from cover to cover. Some riflemen could provide suppressing fire from cover while other units advanced. The manual recommended one group advance from the left flank while the other "keeps up a hot fire on the enemy" from the right.[12] General Pershing discussed fire-and-maneuver as a safer alternative to frontal assaults in his *Tactical Note Number 7*. The commander-in-chief's *Combat Instructions* identified fire-and-maneuver as a key component of his open warfare concept.[13]

Infiltration Tactics

Armies on the western front started experimenting with infiltration tactics in 1915. The purpose of infiltration was to take advantage of the successes of an attack and sidestep the failures. The British Army made effective use of infiltration in 1917 at Vimy Ridge and Passchendaele. They launched assaults against weak points in the German lines and used reserves to exploit the breakthroughs. By 1917, French assault tactics called for platoons to advance in waves. The first wave contained grenadiers and automatic riflemen assigned to neutralize enemy defensive positions and create breaches. Riflemen in the second wave were supposed to penetrate the gaps created in the enemy line by the first wave. The third attack wave was for consolidating gains. In *Manual of the Chief of Platoon of Infantry*, the French warned against assault waves stacking up against areas of persistent enemy resistance. When an enemy machine-gun position stalled elements of the initial wave, the manual instructed the second wave to bypass that position and continue advancing through weaker points in the enemy line.[14]

Major Wilhelm Rohr and General Oskar von Hutier pioneered infiltration tactics in the Imperial German Army. Rohr trained and commanded a "storm" battalion in the German Fifth Army during the Battle of Verdun in 1916. After the success of Roher's "storm" battalion, the German Army established a one-week training program in infiltration for all battalion commanders the following year. Troops of Hutier's Eighth Army bypassed and encircled the Russian defenders of Riga in September 1917. Mobile units broke through weak points in the Russian lines, attack units exploited these breaches, and fortification units consolidated territorial gains. The Imperial German Army published its first manual on "storm trooper" tactics in January 1918. Hutier's German "storm trooper" divisions advanced forty miles during General Ludendorff's spring 1918 offensive.

Infiltration tactics appeared in U.S. Army doctrine during the Great War as well. *Instructions for the Offensive Combat of Small Units* advised first assault waves to avoid pausing to fire their rifles, clear obstacles, occupy enemy trenches, or consolidate territorial gains. These tasks belonged to the second and third waves. The second wave provided covering fire and the third wave secured captured ground. The manual discouraged launching repeated attacks against stubborn enemy positions, instead, it provided guidelines for blending fire-and-maneuver with infiltration. It instructed platoons facing strong resistance to find cover, apply suppressing fire, and wait for reinforcements. Reinforcements had to flank the suppressed strong points, rather than buttress the preceding platoons.[15]

General Pershing witnessed the Imperial German Army's "storm trooper" tactics during the Ludendorff Offensive and factored infiltration into his open warfare concept by late 1918. In *Tactical Note Number 7*, the commander-in-chief encouraged his first assault wave units to circumvent enemy strong points and leave them for subsequent

II. AEF Open Warfare Doctrine

waves. Exploiting breaches as rapidly as possible would prevent the enemy from regrouping. Pershing's *Combat Instructions* urged the second assault waves to avoid stacking behind units that were stalled by enemy resistance. Instead, the second assault waves needed to pass through gaps in the enemy line and flank the persistent strong points.[16] The AEF's recognition of the importance of infiltration tactics was clear as any army on the western front by the time of the armistice. For example, the 131st Infantry Regiment of the Thirty-Third Division utilized "squad rushes" and "leap frog" tactics to traverse the heavily defended wooded terrine of Bois d'Harville during the last two days of the war.[17]

Initiative and Improvisation

American officers and noncommissioned officers inherited a great deal of command responsibility on the battlefield when the U.S. Army embraced Upton's plan for dividing mass linear formations into smaller units. The battlefield independence of Prussian junior officers impressed Upton during his visit to Europe in the mid–1870s. Upton returned to the states and advocated that American officers and noncommissioned officers be permitted to take more initiative in combat.[18] AEF doctrine followed Upton's theory by instructing its battalion, company, and platoon commanders to be aggressive, innovative, and aware of finite strategic goals.

Infantry Drill Regulations United States Army 1911 outlined an infantry commander's responsibilities on the battlefield and identified ideal leadership traits. The manual recognized that controlling infantry formations in combat required a fine sense of timing. Deploying soldiers too soon would result in disorganized firing lines and assault columns. Keeping soldiers in formation too long would attract enemy fire and decrease the odds of survival. Officers were responsible for maintaining order and controlling fire despite this challenge. Furthermore, commanders had to adjust their tactics as battlefield circumstance evolved. The publication defined this as an officer's "tactical instinct." "Tactical instinct" included the ability to order an attack the moment fire superiority was achieved and resist the impulse to pursue retreating enemy soldiers after achieving the immediate objective.

American Tactical Advancement in World War I

Manual for Noncommissioned Officers and Privates of Infantry of the Army of the United States, 1917 urged commanders to stop to consolidate their gains after breaking an enemy position and never sustain a disorderly advance merely for the sake of gaining extra ground.[19]

Instructions for the Training of Platoons for Offensive Action reminded officers to adhere to the basic tactical principles of reconnaissance, surprise, flanking, and gathering intelligence. The manual urged platoon leaders to be mindful of limited objectives, such as enemy strong points, command centers, crossroads, high ground, and razed structures. Selecting limited objectives required skill, as choosing the wrong objectives could cause the larger campaign to fail. *Instructions for the Offensive Combat of Small Units* instructed commanders to pause, regroup, and reassess when moving from one limited objective to the next.[20]

General Pershing's open warfare concept encouraged officers to be aggressive, but make informed decisions. *Tactical Note Number 7* recommended that commanders send out patrols to identify enemy strong points before launching offensives. Officers had to utilize this information to plan their assaults and then execute without hesitation. Pershing advised officers to frequently send situation reports to their superiors so that subsequent assault waves could be directed accordingly. The commander-in-chief reemphasized the importance of using scouting patrols to gather intelligence in his *Combat Instructions*. He expected commanders to use this information to seize the initiative. Pershing, like Upton in the late 1800s, expected his officers and non-commissioned officers to seize the initiative without being reckless with the lives of their men. By comparison, the French Army did not encourage taking the initiative at lower levels of command. In the French *Manual of the Chief of Platoon of Infantry*, choosing assault objectives was left only to battalion commanders.[21] Unlike AEF doctrine, this level of responsibility did not extend to company and platoon leaders.

Open Warfare with Combined Arms

Through four bloody years of trial and error, armies on the western front fused new infantry tactics with modern industrial weaponry in

II. AEF Open Warfare Doctrine

their attempts to create and exploit breakthroughs. At Verdun, the French Army utilized creeping artillery barrages and focused their attacks on limited strategic objectives to recapture one fortress at a time from the Germans in 1916. The British Army combined infiltration tactics with aerial reconnaissance and counter-battery artillery fire at Passchendaele. German "storm trooper" units used machine guns, grenades, and flamethrowers during the Ludendorff Offensive. The French and British counter attacks in summer 1918 saw infiltration tactics combined with tanks and air support for ground forces.

The AEF gradually connected General Pershing's open warfare concept with combined arms throughout 1917 and 1918. Grenades, automatic rifles, machine guns, mortars, special weapons, and chemical agents eventually took their place in flanking enemy strong points, attacking from cover, and conducting fire-and-maneuver. *Notes on Grenade Warfare* prescribed small formations, fire-and-maneuver, and the use of cover for grenadiers. While attacking, grenadiers needed to disperse in groups of two and three to avoid drawing enemy fire. Riflemen had to suppress the enemy while grenadiers flanked. Grenadiers encountering stubborn resistance could entrench and throw grenades from cover. The manual encouraged commanders to utilize aerial reconnaissance to plan grenadier assaults.[22]

Instructions for the Training of Platoons for Offensive Action described automatic riflemen providing suppressing fire from cover while attackers flanked and closed with the bayonet. Automatic riflemen could use marching fire to participate in flanking maneuvers as well. *Manual of the Automatic Rifle* depicted the automatic rifle as a weapon for both fire and maneuver. Automatic riflemen could suppress enemy strong points while assault troops lobbed grenades, fired Stokes mortars, and flanked with the bayonet. The manual included automatic rifles in flank assaults as well. The publication noted that German machine-gun mounts were limited to a thirty-degree field of fire. This made it impossible for enemy machine-gun crews to cover both flanks without moving the weapon and its mount. The manual, therefore, recommended automatic riflemen to converge on enemy machine-gun positions from two flanks simultaneously. This would allow one of the automatic riflemen to approach the enemy position unopposed. When enemy defenders became dislodged, automatic riflemen could use marching fire to pursue the retreating enemy and prevent them from regrouping.[23]

Instructions for the Offensive Combat of Small Units recommended machine guns for long range suppressing fire during assaults. *Notes on the Employment of Machine Guns* claimed that the machine gun's role in an attack was to prevent displaced enemy soldiers from forming new defensive positions. Machine-gun crews had to keep pace with the advancing infantry to perform this task. *Employment of Machine Guns* asserted that machine guns were capable of providing continuous covering fire for infantry attacks so long as machine-gun crews advance behind the offensive. The manual described machine gunners suppressing enemy machine guns from a distance while assault troops flanked with automatic rifles, mortars, and thirty-seven-millimeter cannons.[24]

In *Tactical Note Number 7*, General Pershing stated that machine guns, mortars, and thirty-seven-millimeter cannons were ideal additions to his open warfare. The commander-in-chief's *Combat Instructions* urged commanders to use automatic rifles, mortars, and thirty-seven-millimeter guns to suppress enemy machine gunners during flank attacks. Pershing instructed assault troops to use grenades while flanking enemy machine-gun positions. *Gas Manual, Part II* even prescribed mustard gas for suppressing strong points while the infantry infiltrated enemy lines.[25] By the end of the Great War, U.S. Army doctrine started bringing the tactical concepts of Pershing's open warfare and combined arms together. The U.S. Army continued to develop tactics for breakthrough and exploitation through suppression and maneuver with modern industrial weaponry during the interwar period and the Second World War.

Conclusion

Like the French, British, and Imperial German armies, U.S. Army doctrine included small irregular formations, fire-and-maneuver, use of cover, infiltration, and battlefield improvisation by late 1918. Assault columns, skirmish lines, and small flexible formations supplanted costly

II. AEF Open Warfare Doctrine

mass linear frontal attacks. AEF publications encouraged the employment of disciplined suppressing fire and "shock" when closing with the enemy. American infiltration tactics called for soldiers in the first assault wave to bypass enemy strong points and conduct flank attacks. Pamphlets and manuals urged AEF commanders to use reconnaissance and surprise to seize the initiative, improvise tactics, and concentrate on capturing limited objectives. Commanders knowing when to be aggressive and when to persevere was critical to overall battlefield survival and success. Some American tactical pamphlets and manuals even linked these techniques with combined arms. The combined use of grenades, automatic rifles, machine guns, mortars, special weapons, and chemical agents complimented General Pershing's open warfare concept. By November 1918, Pershing expanded Uptonian survival doctrine, observed the advanced infantry tactics of other armies on the western front, and provided the AEF with his own brand of open warfare. The core principles of Pershing's open warfare continued to serve as the foundation of U.S. Army infantry assault tactics throughout the twentieth century and into the twenty-first.

III

AEF Combined Arms and Open Warfare Training

While the AEF incorporated combined arms and General Pershing's open warfare tactics into its published doctrine, it also attempted to merge these new concepts into its training regimen. Pershing set aside one month of advanced tactical instruction for each American division arriving in France even before his open warfare concept was published in September 1918. The commander-in-chief also removed large numbers of officers and noncommissioned officers from the front lines to undergo additional tactical training so they could return to their units and teach these new tactics to the troops under their command. In the spring of 1918, however, Pershing determined that the need for manpower to blunt the German offensive on the western front superseded thorough training. Weapons shortages also hampered the AEF's ability to conduct extensive combined arms and open warfare instruction. Commander of the Twenty-Seventh Division Major General John F. O'Ryan wrote in 1921, "What handicapped the division and the American army ... was ... the problem of equipping them the material things needed ... for their effective training."[1] Finally, most sections of the AEF did not develop standardized combined arms and open warfare training methods until late 1918. The armistice went into effect before the trainees receiving this new instruction reached the western front. Despite these shortcomings, training did make a modest contri-

III. AEF Combined Arms and Open Warfare Training

bution to the AEF's ability to conduct combined arms and open warfare by the final phase of the Meuse-Argonne Offensive.

Pre-War Training and Organization

The U.S. Army spent the majority of the latter half of the nineteenth century engaged in unconventional warfare against Native Americans. Marksmanship and small-unit tactics, with the occasional combined use of cavalry, infantry, and artillery, proved useful in these campaigns. During this time, U.S. Army instruction consisted of garrison training, such as marching drill, calisthenics, and classroom study. Lieutenant General William T. Sherman argued in favor of larger field maneuvers during his stint as commanding general from 1869 until 1883. All combat arms finally conformed to Sherman's proposal when the army conducted its first field-training maneuvers in the 1890s. The Regular Army, however, was too small and spread out across the American West to conduct large-scale maneuvers. Consequently, the United States sent an army of undertrained regulars and inexperienced volunteers into combat with disastrous results during the Spanish-American War. American troops suffered unnecessary losses launching frontal assaults against fortified positions and Spanish Maxim machine guns. The U.S. Army owed its victory more to a twelve-to-one numerical advantage over the Spanish Army than its tactical prowess. The United States might not have emerged as an imperial global power in 1898 had its adversary been an army other than Spain's.

On the eve of America's entry into the Great War, basic U.S. Army training involved six months of garrison training and six months of field instruction. The garrison training program reflected the army's continued preoccupation with marching drill, calisthenics, and some classroom instruction. The field-training regimen represented the army's gradual adoption of General Sherman's combat-simulation training concept. The reforms of Secretary of War Elihu Root from 1899 until 1904 established this program of instruction as the standard across the service. This was the first time the U.S. Army imposed uniform training guidelines on all American soldiers, including regulars and volunteers. Furthermore, the "Root Reforms" instigated the formation of specialized artillery schools at Fort Sill, Oklahoma and Fort Monroe, Virginia in 1902. The reforms also increased the size of the army from about twenty-five thousand in 1898 to over one hundred thousand by 1915.[2]

Members of the British Army general staff observed U.S. Army field-training maneuvers in Connecticut in August 1912. Twenty thousand American soldiers participated in the exercise. Most of these troops were National Guardsmen. The British observation report made note of American advanced infantry tactics, including fire-and-maneuver and the use of cover. The British officers saw a large body of infantrymen advance to within six hundred yards of a defensive position and form a firing line. A smaller group of soldiers provided covering fire from a protected position eight hundred yards away. The British noted the Americans' effective use of cover during the mock assault. Few American officers issued orders to fire at will, so most infantry units fired by volley during the maneuvers.

The British report revealed American ignorance of combined arms tactics and the limited availability of modern weaponry. Aside from mounted cavalry patrols providing reconnaissance for infantry commanders, coordination between the infantry and other arms was nonexistent. The Americans used three observation aircraft during the exercise, two Curtises and one Burgess-Wright. One of the Curtis aircraft was on loan from a private owner. Artillery commanders made some use of telephones to direct battery fire, and there was some wireless communication between the observation aircraft and the artillery. The artillery batteries only employed direct fire on visible targets, however, so their reliance upon aerial observation was minimal. Automobiles and motorcycles carried messages behind the lines, but remained outside the combat zone. Horse-drawn wagons provided the majority of logistical support.

The U.S. Army's Connecticut Maneuvers were smaller and involved less advanced weaponry than most other army exercises the British observed in 1912. The French and German armies each conducted maneuvers with two full army groups. Their training exercises featured motorized infantry, machine guns, and extensive aerial reconnaissance. The French employed sixty aircraft and the Germans utilized thirty-

III. AEF Combined Arms and Open Warfare Training

eight. Tactically, however, the Americans were on par with other armies. The French Army made some use of fire-and-maneuver and cover but still relied heavily on volley fire. Furthermore, despite their advanced weaponry, both the French and Germans failed to effectively coordinate arms.[3]

Wartime Plans and Problems

American soldiers received less than two months of stateside instruction before being sent off to fight in the Spanish-American War. When the United States entered the Great War, the War Department tried to avoid the training inadequacies of 1898 by setting stateside instruction at six months for each division. The AEF's 1917 organization plan called for divisional stateside training to include artillery, chemical agent, machine-gun, mortar, flamethrower, aerial observation, and tank instruction. This was far too ambitious considering the lack of modern industrial weaponry in the United States at the time. A scarcity of modern rifles in the United States meant that many American soldiers did not even fire a weapon until they reached France. During the Thirty-Third Division's stateside training, Lieutenant Colonel Frederic L. Huidekoper observed "a conspicuous dearth of the materiel and supplies which are indispensable to a properly equipped military force." The "Prairie Division" assembled at Camp Logan in September 1917, had no artillery or machine guns for training until November, and then cancelled instruction altogether in January 1918 due to the unexpectedly harsh Texas winter weather that year.[4]

French and British training officers arrived in the United States in 1917 to help with the situation. They conducted grenade, gasmask, machine-gun, and bayonet exercises in mock trenches across the southern and southeastern United States. For example, the Twenty-Seventh Division trained in an enormous eight-mile system of trenches at Camp Wadsworth, South Carolina. General O'Ryan praised the French and British instructors at Camp Wadsworth for playing a key role in preparing his "New York Division" for combat in Europe.[5] Unfortunately, weapons shortages plagued French and British training efforts in the United States as well. The Twenty-Seventh had no thirty-seven-millimeter guns and only a few obsolete Colt machine guns and mortars during their time at Camp Wadsworth from September 1917 until April 1918.[6] Many stateside trainees used baseballs and rocks as grenades, wooden boards as machine guns, and broomsticks as rifles. For the Eighty-Second Division at Camp Gordon, Georgia, "the training program so far outstripped available equipment that regiments ... turned out for drills ... with wooden guns minus all other ordnance." Needless to say, most AEF stateside training consisted of marching drill, calisthenics, and some target practice. The "All American Division" passed many hours at Camp Gordon "road marching."[7] These limitations in stateside training were not entirely unique to the U.S. Army during the Great War. The British Army's stateside regimen in 1914 and 1915 comprised mostly of marching, physical training, marksmanship, and bayonet drill conducted with wooden boards and broom handles.

Other logistical problems delayed AEF stateside training. In keeping with President Woodrow Wilson's pretense at neutrality since 1914, the U.S. Army did not prepare for war prior to the United States' declaration of war on Imperial Germany on 6 April 1917. This left the army without adequate training facilities when the United States finally entered the conflict. The War Department ordered the makeshift construction of thirty-two training camps between June and September 1917. Most of these were nothing more than a field of wooden shacks and canvas tents. The War Department intended each camp to accommodate forty thousand trainees. The population of most camps far exceeded this number and they quickly became overcrowded. Once again, these deficiencies were not entirely unique to the AEF. British trainees in 1914 and 1915 lived in tents, huts, and some even returned home for a meal and a bed every evening.[8]

Stateside AEF divisional training also suffered when the War Department periodically removed experienced officers and enlisted men from existing regiments and dispersed them among newly formed units. This was particularly detrimental to National Guard divisions, such as the Twenty-Seventh and Thirty-Third, whose troops had the benefit of training together prior to the war.[9] Colonel Huidekoper reflected on the War Department's "readjustments" of the Thirty-Third as "not conduced to systematic or uninterrupted training of the

III. AEF Combined Arms and Open Warfare Training

troops."[10] *The History of the Seventy-Ninth Division, AEF* recounted that "The War Department made ... a long series of drafts ... robbing the Division of many of its trained men at a time when they were most needed."[11] *The Official History of the 82nd Division, AEF* described the loss: "Everyone felt that infanticide was being committed upon newly born regiments."[12]

AEF divisions continued training when they reached France. The infantry trained in twenty-one zones around General Pershing's headquarters at Chaumont. American artillery units trained separately from their infantry divisions in LeValdahon. Pershing established a three-month training program for each newly-arrived division. He originally wanted five months of instruction, but the Allies and the War Department favored getting American troops to the front faster. The AEF initially relied on French and British training officers until enough American instructors arrived in France. The French and British offered each American division one month of advanced weapons instruction. The French gave machine-gun, grenade, automatic-rifle, and thirty-seven-millimeter-gun lessons. The British taught mortars, sniping, scouting, the bayonet, and marksmanship. AEF divisions then spent a month in a quiet sector at the front. This presented them with an opportunity to gain some combat experience conducting reconnaissance patrols and trench raids. Pershing reserved the last month of training for his open warfare tactics, even before clarifying exactly what those techniques were.

When Pershing finally defined his open warfare concept in the fall of 1918, he ordered division commanders to pull officers and noncommissioned officers off the line so they could learn these new tactics and teach them to the soldiers under their command. The commander-in-chief also ordered one division from each corps rotated off the line for advanced tactical training. Many of these "retrained" American officers and enlisted men returned to the front in time for the final phase of the Meuse-Argonne Offensive. For this reason, the AEF was better trained on 1 November than at any earlier point in the war.

Unfortunately, these efforts to teach the AEF combined arms and open warfare proved both too little and too late. When the German offensive in the spring of 1918 broke through Allied lines, Pershing scrambled to meet the demand for troops at the front. The commander-in-chief eliminated the third month of his divisional training program in the summer of 1918. This put an end to the expectation that all AEF divisions would receive open warfare instruction before entering combat. In fact, the First Division was the only AEF division to complete all three months of Pershing's originally prescribed training program. In accordance with Pershing's call for manpower in 1918, American schools providing advanced training for officers and enlisted men pulled from the front abbreviated their programs as well. Stateside training was also shortened in response to the need for troops on the western front. A draftee who enlisted in Philadelphia in July 1918 would find himself in combat in France in October of that same year. The War Department even ordered the U.S. Military Academy at West Point and the Leavenworth schools to abridge their curricula.

Pershing Versus the Allies

The AEF's commander-in-chief harbored reservations about the War Department's decision to invite French and British training officers to the United States. He feared too much European training would condemn the AEF to the same trench-warfare tactics that turned the western front into a stalemate after 1914. General Pershing believed his open warfare, although not yet clearly defined, would restore mobility to the battlefield and drive the Germans out of Flanders and France. The commander-in-chief begrudgingly agreed to allow the French and British to insert one month of their instruction into his three-month training regimen. The primary reason why he conceded to this was because there were not enough American instructors in Europe at the time. Pershing immediately took steps to undo this. He telegraphed the War Department in December 1917 to suggest that one third of every division's officers and noncommissioned officers be sent to France five weeks ahead of the rest of their comrades for training. These trained officers would then instruct the remainder of the division when it arrived in France.[13] Through this process, Pershing eventually replaced most French and British training officers with Americans by the war's end.

III. AEF Combined Arms and Open Warfare Training

American concerns about French and British methods of instruction were not limited to General Pershing. Colonel Harold B. Fiske became Chief of the Training Section of the General Staff, General Headquarters, in February 1918. He criticized Allied training in a 4 July 1918 memorandum to AEF Chief of Staff, Major General James W. McAndrew. Fiske accused French instructors of teaching infantrymen to rely on artillery support to protect them across the battlefield, rather than depending upon their own skill with the rifle and bayonet. Fiske asserted, "The French do not like the rifle, do not know how to use it, and their infantry consequently is too entirely dependent upon a powerful artillery support." He complained that French and British training was passive, defensive-minded, and devoid of officer initiative. Fiske criticized both the French and British: "Their infantry lacks aggressiveness and discipline. The British infantry lacks initiative and resource."[14] Major General James G. Harbord's *Final Report of Assistant Chief of Staff, G-5* issued on 30 June 1919 expressed postwar contempt for French and British instruction. Harbord served as AEF Chief of Staff from May 1917 until May 1918. He accused Allied training of programming soldiers for blind obedience, rather than providing them with an understanding of the purpose behind combat orders.[15]

Although Pershing condemned the French and British for neglecting his open warfare concept, he appreciated their advances in combined arms. He urged his officers to take advantage of opportunities to learn combined arms tactics from the French and British in his *General Orders Number 29: Instructions for Officers Visiting the French or British Lines or Serving with French, British, or American Units at the Front*. The 14 February 1918 directive instructed AEF officers to observe the French and British, take notes on their use of combined arms, and submit their findings to general headquarters. American officers visiting the French and British combat sectors took account of their combined use of infantry, artillery, machine guns, automatic rifles, grenades, rifle grenades, mortars, thirty-seven-millimeter guns, chemical agents, tanks, and aircraft.[16] This helped the AEF to realize the importance of combined arms and incorporate them in its own tactical development.

American Tactical Advancement in World War I

Pershing Versus the War Department

General Pershing's training expectations also conflicted with those of the War Department, especially in terms of the level of instruction received by AEF divisions before they departed for Europe. The commander-in-chief often attributed the poor fighting quality of American troops in France to their inadequate stateside training. Pershing sent a cable to AEF Adjutant General, Brigadier General Benjamin Alvord, Jr., on 22 December 1917 criticizing the aptitude of newly-arrived officers. He complained that American officers came to Europe with poor marksmanship skills and little knowledge of advanced tactics. Furthermore, he noted their inability to pass on their limited knowledge to the soldiers under their command. The commander-in-chief expected division commanders to be prepared to teach marksmanship and tactics to their troops by the time they arrived in France. Pershing addressed this problem at the U.S. Army infantry and artillery schools at Fort Sill by ordering the Director of Training at Fort Sill, Major General John F. Morrison, to dismiss all instructors who did not adhere to Pershing's training expectations.[17]

The commander-in-chief aired his concerns about the AEF's stateside instruction again in the spring of 1918. On 18 April, Pershing complained to the adjutant general about American officers and enlisted men still arriving in France with insufficient tactical knowledge. He noted marksmanship and combat-discipline problems as well. The commander-in-chief called for increased stateside combined arms training in infantry/artillery coordination. He ordered improved marksmanship skills by setting the U.S. Army rifle proficiency standard at six hundred yards, as the average American soldier in France in 1918 was accurate at only three hundred yards. Pershing expressed the need to instill combat discipline in junior and noncommissioned officers by teaching them to instinctively assume command if their superiors fell in battle. A cablegram from the commander-in-chief to the U.S. Army's Chief of Staff, General Peyton C. March, on 24 April 1918, suggested divisions remain in the United States until they received at least four months of instruction.[18]

General March addressed Pershing's comments on 7 May and 16

III. AEF Combined Arms and Open Warfare Training

June 1918. He telegraphed the commander-in-chief to assure him that all trainees in the United States received rifle instruction beyond three hundred yards. March explained how British and French demands for American divisions in Europe made it necessary for the War Department to reduce the duration of stateside instruction to less than three months. The chief of staff recommended Pershing lengthen AEF division training in France to compensate for the time lost stateside. This was an unrealistic solution, as the commander-in-chief faced even greater pressure from the Allies in Europe to rush his troops to the front. Frustrated, General Pershing directed his next cablegram over March's head to Secretary of War Newton D. Baker. This placed strains on the relationship between the commander-in-chief and the chief of staff, which resulted in March's public criticism of Pershing's leadership after the war. The commander-in-chief complained to Baker that deficient marksmanship and tactical training in the United States doomed the AEF to the same stagnant trench-warfare the French and British had experienced since 1914. Secretary Baker offered no response.[19]

General March defended AEF stateside training in his memoirs after the war. He asserted that the decision to shorten training and rush American divisions into combat in 1918 prevented the war from carrying into 1919. March claimed that, despite the lack of formal instruction, American soldiers represented the finest troops on the western front. He believed the United States' democratic tradition and capitalist work ethic provided Americans with a cultural predisposition for military service. He considered U.S. Army training facilities, such as West Point and Fort Leavenworth, equal to any European military institution. March insisted that the AEF was as tactically sound as any European army in 1918, in better physical condition, and free from the stagnant mentality of trench warfare.[20]

Standardization

By the outbreak of the Great War, the U.S. Army was only beginning to reach a consensus on standard, service-wide, tactical doctrine and training. General O'Ryan of the Twenty-Seventh Division described the mixed and conflicting ideas about the goal of training across the service in 1917: "numerous officers ... believed that the war would be won with the bayonet.... Others ... advocated ... the development of machine gunners.... Still others ... grenades."[21] The AEF finally established a process for standardizing instruction across the service by summer 1917. Lieutenant Colonel Paul B. Malone became Chief of the Training Section of the General Staff, General Headquarters, in August 1917. He and General Harbord immediately established a means for passing tactical doctrine and training guidelines from the highest echelons of AEF command to the smallest fighting units. Army schools drafted doctrine and passed it to corps schools for dissemination among divisions. This created a deluge of tactical and training literature, not all of which was useful. O'Ryan asserted that the Twenty-Seventh Division's training "was complicated by a veritable avalanche of books, booklets, pamphlets and bulletins covering every phase and aspect of the conduct of war. Some ... were illuminating.... Many obsolete."[22] Division commanders sent their officers to corps schools to learn new doctrine. These officers returned to their units to instruct soldiers under their command. Division commanders supervised divisional training to insure uniformity.[23]

Corps Schools

Corps schools played a major role in General Pershing's plan to provide advanced tactical training to American officers and noncommissioned officers in France. The AEF's three corps schools graduated a total of 13,916 officers and 21,330 noncommissioned officers during the Great War. Initially, division commanders were hesitant to send their experienced officers and NCOs to the corps schools. They could not spare competent leadership at the front, and corps schools often retained their best graduates as instructors. Nevertheless, many commanders soon recognized the value of the retrained officers who returned from the corps schools four weeks later. Publications pertaining to all three corps schools reflected the AEF's effort to expose its trainees to combined arms tactics and Pershing's open warfare concept. These reports also reveal how the demand for manpower at the front in 1918 compromised this advanced training initiative, as each

III. AEF Combined Arms and Open Warfare Training

school abbreviated its four-week program to meet the call for frontline troops.

General Harbord's 30 August 1917 *School Project for the American Expeditionary Forces* outlined the function of AEF corps schools. AEF Corps schools provided infantry, artillery, engineer, cavalry, chemical agent, signal, aviation and observation balloon training to majors and lieutenant colonels pulled from the front. Harbord's publication illustrated the schools' attempts to provide specialized instruction in a variety of weapons and techniques essential to combined arms. The infantry schools handled machine-gun, mortar, automatic rifle, grenade, and tank instruction. The engineer schools taught road building, bridge construction, and artillery ranging. The signal schools consisted of telegraph, telephone, radio, pigeon, and semaphore.[24]

I Corps School opened on 15 October 1917 in Gondrecourt, France. I Corps School submitted its final report to General Pershing on 11 November 1918. The report provided evidence of the school's combined arms instruction. The infantry school's regimen included the combined use of the rifle and bayonet, grenade, automatic rifle, machine gun, mortar, and thirty-seven-millimeter gun when attacking. The artillery school taught 75-millimeter cannon and 155-millimeter howitzer crew commanders to coordinate with officers of the infantry and air service. I Corps School graduated 12,535 officers and noncommissioned officers during the Great War.[25]

II Corps School opened on 4 February 1918 in Chatillon-sur-Seine near Langres, France. Colonel H. L. Cooper submitted the *II Corps School Report* on 21 April 1919. He asserted that the AEF's publication of Pershing's open warfare doctrine occurred too late to produce battlefield results during the Meuse-Argonne Offensive. For example, II Corps School did not begin open warfare instruction until after the start of the Meuse-Argonne Offensive. Colonel Cooper claimed that most AEF officers and noncommissioned officers did not have enough open warfare training to seize the initiative and improvise when attacking the strong German defenses in the Argonne Forest. Furthermore, many officers were not confident or knowledgeable enough to incorporate advanced weapons, such as mortars and thirty-seven-millimeter guns, into their assaults.[26]

III Corps School opened in Clamecy, France, on 2 September 1918. The school issued its final report on 9 April 1919. The report provided evidence of the school's combined arms efforts, including instruction in the rifle and bayonet, automatic rifle, machine gun, grenade, mortar, and thirty-seven-millimeter gun, but showed little sign of Pershing's combined arms. III Corps School graduated 7,675 officers and noncommissioned officers during the war.[27]

The final report of Inspector General of AEF General Headquarters, Major General Andre W. Brewster, presented inspecting officers' overall assessment of the effectiveness of the AEF's corps school programs during the war. Inspection officers visited one division in each corps every day during the war. These inspectors reported the division's compliance with AEF tactical doctrine back to General Pershing. General Brewster's final report attested to the benefit of removing American officers and noncommissioned officers from the front for advanced tactical training at the corps schools. His report confirmed that most inspectors noted improved tactical performance after a division's officers and noncommissioned returned from training at the corps school.[28] As a result of the corps schools, the AEF was better led during the last phase of the Meuse-Argonne Offensive than any other point in the war.

Other Schools

The U.S. Army's officer candidate schools, school of the line, and staff college also played a role in exposing AEF leadership to combined arms and open warfare training. The National Defense Act of 1916 established America's first officer candidate schools. These ninety-day programs became increasingly important to the AEF as officer casualties mounted, especially among lieutenants commanding platoons. These schools graduated 6,895 infantry officers, 3,393 artillery officers, 1,332 engineer officers, and 365 signal officers during the Great War. The AEF started an officer candidate school in Langres on 20 December 1917. The school's commander, Lieutenant Colonel S. L. Pike, issued the Army Candidates School report on 15 February 1919. The report provided evidence of combined arms and Pershing's open warfare in

III. AEF Combined Arms and Open Warfare Training

the school's curriculum, including advanced weaponry, combined arms tactics, and fire-and-maneuver. By 11 November 1918, 3,242 second lieutenants graduated from the school.[29]

The U.S. Army School of the Line at Fort Leavenworth offered a three-month program of lectures, written problem solving, and tactical exercises in combined arms and Pershing's open warfare concept. The school graduated 497 field officers during the war.[30] The "Root Reforms" established the U.S. Army's General Staff College at Fort Leavenworth just prior to the Great War. The college graduated about seven hundred officers by 1917. The fact that these prewar graduates dominated General Pershing's staff bred animosity among officers outside the so-called "Leavenworth Crowd." The War Department temporarily closed the stateside general staff college during the Great War due to low enrollment, so the AEF opened its own staff college at Langres in October 1917. Brigadier General Alfred W. Bjornstad served as the college's director. The school offered a three-week course in advanced weapons, tactics, and written instruction in battlefield problem solving for division commanders, corps commanders, and staff officers rotated from the front. The college graduated 534 officers during the war.[31]

Adjustments and Limitations

On 5 March 1918, General Pershing's *General Orders Number 35* redefined the scope of AEF corps school instruction to make combined arms first and foremost. The orders arranged for a mandatory four weeks of instruction for officers and noncommissioned officers of all AEF divisions. Pershing called for training exercises involving automatic rifle, grenade, machine-gun, mortar, and thirty-seven-millimeter gun support for the infantry. He advocated for war games designed to teach coordination between the infantry and artillery. Engineer training focused on supporting the infantry's advance by removing obstacles and building and repairing roads and bridges. Aviation and observation balloon instruction concentrated on directing artillery fire.[32] Unfortunately, time constraints imposed by the need for manpower on the battlefield in summer 1918 precluded most of these plans. Nevertheless,

Pershing's orders revealed his appreciation of the importance of more sophisticated tactical learning by early 1918.

In addition to the curtailment of training periods in the summer of 1918, carelessness and sloth plagued AEF instruction as well. On 24 June 1918, the commander-in-chief reported the findings of General Brewster in *Bulletin Number 41*. Brewster's training inspectors discovered American training officers who neglected to correct trainee mistakes. He wrote, "It has been observed that in many training areas there is a lack of that seriousness and vigor on the part of the instructors that is so necessary.... Mistakes are often passed over without notice." Furthermore, few instructors actually participated in training exercises alongside their students. This robbed trainees of the opportunity to learn by example. Pershing and Brewster also found that many American officers were not taking advantage of opportunities to instruct their troops at the front. The commander-in-chief's 7 July 1918 *Bulletin Number 44* reminded commanders to use any and all downtime at the front to conduct training exercises; even if it meant setting aside administrative business and paperwork.[33] Lulls at the front offered many American soldiers their only opportunities to supplement the hasty training they had received before entering combat.

Postwar recollections also cited problems with AEF instruction. After the war, General Harbord asserted that abbreviated training prevented the AEF from reaching its full potential during the Great War. He recalled how few American officers completed their prescribed three months of instruction and most AEF schools shortened their programs to six weeks or less. General Fiske concluded that "this war has not reversed all the lessons of the past by proving that tacticians can be made in a few months training or ... resourceful divisions can be made by a few maneuvers." His postwar reflections cited AEF battlefield deficiencies in artillery support for the infantry, infiltration tactics, fire-and-maneuver, and the use of cover. Even General Pershing's published postwar memoir conceded that the Meuse-Argonne Offensive may have gotten off to a better start had the AEF been more thoroughly trained.[34]

III. AEF Combined Arms and Open Warfare Training

Artillery Training

Postwar reports revealed the artillery's efforts to promote combined arms through coordinated training with other arms and motorization. These reports also illustrated the factors that limited the artillery's combined arms progress, such as weapons shortages in the United States. Colonel Huidekoper asserted that the Thirty-Third Division's artillery brigade had not "fired so much as one single round in target practice" before departing for Europe.[35]

The *Final Report of the Chief of Artillery, American Expeditionary Forces*, by Major General Ernest Hinds, outlined AEF artillery instruction in France. AEF artillery training in Europe entailed four months of instruction. The first included weapons operation and signaling. The second involved service in a quiet sector of the front. For the third month, artillery brigades joined their assigned infantry divisions for combined arms training. The chief of artillery reserved the fourth month for artillery commanders and specialists. Like the infantry, the demand for manpower at the front in the summer of 1918 interfered with artillery instruction. Most American artillery training centers in France shortened their programs from four months to a month-and-a-half by fall 1918. As a result, less than half of the AEF's twenty-two artillery brigades progressed beyond the first month of the prescribed training regimen, and only a handful of those brigades participated in combined arms training with their assigned infantry divisions. Those few divisions fortunate enough to train with their artillery brigades experienced positive results in combat. For example, General O'Ryan affirmed that "there is no question … the 27th Division, as a result of their rather radical training … were very much more effective … when they went forward behind a supporting barrage."[36]

AEF artillery schools worked to improve the artillery's combined arms capabilities by enabling artillery crews to keep pace with the infantry through motorization. The AEF's heavy artillery training center in Vincennes, France opened a tractor school in May 1918 to teach heavy artillery crews how to use tractors to tow their howitzers into battle. The AEF had almost nine hundred tractors for towing heavy artillery when the war ended. A motor transport school with the same goal for AEF field artillery opened in St. Maur in November 1918. Its trainees learned to drive tractors, automobiles, trucks, and motorcycles. Despite these efforts, AEF artillery still relied primarily upon horses during the Great War.

Colonel Adam F. Casad's *Ordnance Report* of December 1918 quantified the AEF's wartime artillery capabilities. The U.S. Army possessed less than one thousand obsolete artillery pieces before entering the Great War in April 1917. According to the report, the AEF had 1,862 field artillery pieces by the armistice. Despite this growth, there were mixed opinions as to whether or not the AEF had adequate artillery support during the war. General Pershing and the majority of his staff believed that the AEF always had enough artillery. Major General Charles Summerall, one of the commander-in-chief's senior subordinates, disagreed. Summerall was the AEF's most ardent proponent of artillery support for infantry. He claimed the AEF never had more than half the artillery required to support its forty-two divisions during the war.[37]

The Chief of the Statistics Branch of the General Staff Colonel Leonard P. Ayres completed *The War with Germany: A Statistical Summary* in August 1919. This publication disputed General Summerall's claim. Ayres asserted that there was sufficient artillery support for all AEF divisions in combat at all times during the war. According to Colonel Ayres, the AEF employed a total of 2,250 field artillery pieces during the Great War. He admitted that some infantry divisions entered combat without artillery support, but these incidents were the result of sluggish artillery training, not a lack of guns. Ayres claimed that the limited availability of artillery for training purposes posed a more serious problem. AEF training centers in France experienced a shortage of light artillery pieces from September 1917 until September 1918. The need for heavy artillery at AEF training facilities in Europe was even more severe and lasted the entire war.[38]

General Hinds' report identified some additional shortcomings of AEF artillery training. He claimed that American artillery needed to be totally motorized to serve as an effective combined arms component. Furthermore, Hinds argued that combined arms needed to be improved through increased coordinated instruction between artillery,

III. AEF Combined Arms and Open Warfare Training

infantry, and air service officers. He stated, "There should be a greater amount of time devoted to the combined training of our infantry and field artillery, with interchange of officers of these two arms for a few months period of training. The liaison between the artillery and the air service must be improved." He believed that aerial observers lacked adequate artillery training and were unfamiliar with the functionality and specifications of the weapons for which they gathered targeting data. General Hinds claimed that aerial observation squadrons should have been permanently attached to artillery brigades to increase their familiarity with artillery operations.[39]

Colonel Harold Fiske claimed that American artillery schools established combined arms programs too late in the war. He observed, "A very successful field artillery school ... was established at LeValdahon about the time of the armistice ... stress was laid upon close support of the infantry, to facilitate work with which a fine battalion of infantry was assigned to the school." Colonel M. E. Locke submitted the *Center of Artillery Studies Report* in July 1919. The center planned a three-week combined arms course with artillery, infantry, air service, and engineer officers training together. Unfortunately, General Pershing's demand for officers at the front left the school understaffed and unable to begin instruction until after 11 November 1918. Realizing that artillery coordination with other arms would play a major role in future wars, Colonel Locke recommended that the school be allowed to follow through with its intended combined arms curriculum in the United States after the war. He claimed, "The establishment of some institution akin to it seems to be very desirable for purposes of study during peace time. As was the case in Europe it is believed the class should be ... officers of field and coast artillery, with a certain representation of other combatant arms of a division, such as the infantry, aviation, engineers, and machine guns."[40] The U.S. Army, therefore, intended to continue the AEF's artillery combined arms training after the Great War.

Chemical Warfare Training

Postwar reports demonstrated the evolution AEF chemical agent training, from defensive to offensive. The U.S. Army addressed combined arms as it explored the offensive potential of chemical agents and its coordination with other arms.

The U.S. Army had not prepared for chemical warfare prior to 1917. The War Department authorized the U.S. Army's first gas regiment in the summer of 1917. The AEF's first gas school opened in Langres, France, in October 1917. Unfortunately, a shipment of faulty gas masks and a lack of Livens projectors and mortars delayed its training exercises. The *Army Gas School Report* of 11 November 1918 indicated that the majority of the school's chemical agent training dealt with defending against chemical attacks. The chemical warfare programs of the AEF's corps schools focused on defense as well. This defensive focus was consistent with the AEF's overall reluctance to use poison gas for fear of German retaliation. General Pershing and his commanders were aware of the offensive applications of chemical agents, but employed them only sparingly against the enemy for much of the war.

The *Activities of the Chemical Warfare Service (July 5, 1917, to March 15, 1919) Report* chronicled the evolution of Pershing and the AEF's attitude toward using chemical agents in battle. Time spent on the western front eventually convinced the commander-in-chief of the offensive potential of poison gas. He finally created an offensive branch within the AEF's chemical warfare service in May 1918. The offensive section's purpose was to deliver poison gas, smoke, and incendiaries *via* artillery, Livens projector, mortar, and grenade. It was responsible for instructing the artillery and infantry in the use of chemical agents as well. On 27 May, Pershing attached chemical officers to all armies, corps, and divisions. These officers worked alongside infantry and artillery commanders to incorporate chemical agents into their battle plans.

General Pershing established an independent AEF Gas Corps on 3 June 1918 and placed Lieutenant Colonel Amos A. Fries in command. The Gas Corps was responsible for developing the AEF's offensive chemical agent capabilities in France. Meanwhile, the War Department formed the Chemical Warfare Service in the United States with Major General William L. Sibert as its chief. The U.S. Army Chemical Warfare Service was in charge of developing chemical munitions and equipment in the United States. On 7 September, just before the Saint Mihiel Offensive, Pershing and the AEF's attitude toward chemical agents had

III. AEF Combined Arms and Open Warfare Training

evolved from strictly defensive to offensive. General Pershing ordered 20 percent of all American artillery shells be filled with poison gas during the Meuse-Argonne Offensive.[41]

Machine-Gun Training

General Pershing's general orders and bulletins represented the AEF's realization of the machine gun as an offensive weapon and combined arm. By late 1917, the commander-in-chief's orders and bulletins encouraged coordination between machine guns and infantry in training and battle planning.

Most AEF divisions had little or no machine-gun training before departing for Europe. For example, the post-war histories of both the Twenty-Seventh and Eighty-Second divisions reported having only a handful of outdated Colt machine guns available to them during their stateside instruction.[42] The AEF's first machine-gun school opened at Fort de Peigney, France, in December 1917 and offered machine-gun instruction to officers and noncommissioned officers. American machine-gun training initially focused on the Hotchkiss and Vickers machine guns as defensives weapons. Soon realizing the offensive application of the machine gun, General Pershing ordered training between machine-gun and infantry units to foster combined arms. His *General Orders Number 82*, dated 28 December 1917, directed machine-gun companies to train with their assigned infantry battalions. The commander-in-chief's 23 May 1918 *Bulletin Number 30* ordered infantry division commanders and machine-gun battalion commanders to coordinate their battle plans. His *General Orders Number 91*, dated 10 June 1918, urged infantry and machine-gun officers to plan combined arms tactics prior to every assault.[43]

Automatic Rifle Training

Postwar reports revealed American automatic-rifle instruction's evolution from basic weapon operation to its function as a combined arm. Unfortunately, weapons shortages deprived the AEF of using the finest automatic rifle of the period.

American Tactical Advancement in World War I

The U.S. Army Infantry Specialist School opened at Chatillon-sur-Seine near Langres in December 1917. Initially, its automatic-rifle program focused on firing the weapon effectively and keeping it mobile in combat. The school eventually adopted combined arms tactics into its training regimen, such as utilizing automatic rifles and grenades during assaults.

AEF automatic-rifle instruction included the French Chauchat and the American-made M1918 Browning automatic rifle or BAR. The *Infantry Specialist School Report* of 11 November 1918 considered the Browning to be the better of the two weapons. The BAR fired at a rate of five hundred rounds per minute. It weighed seventeen pounds and took a twenty-round magazine. BAR training was rare, however, due to the scarcity of this weapon. The U.S. Army did not issue Brownings to divisions at the front until September 1918. Only two AEF divisions carried BARs during the Meuse-Argonne Offensive. This left the tragically inferior Chauchat as the AEF's most widely used automatic rifle. Colonel A.E. Phillips' *Machine Gun and Small Arms Activities Report*, dated 11 November 1918, attributed high casualties among automatic riflemen to frequent Chauchat malfunctions. The weapon's poor design resulted in constant jams. In fact, a considerable degree of American automatic-rifle training dealt with techniques for coping with the Chauchat's malfunctions. The *Infantry Specialist School Report* stated, "The first and fundamental aim in instruction in the automatic rifle section is to impart to the student an expert and thorough knowledge of the mechanical characteristics of the rifle, with particular stress upon the reduction of stoppages." The French *Manual of the Chief of Platoon of Infantry* made the absurd suggestion that soldiers reduce the likelihood of jams by carrying Chauchats in special cases to protect them from dirt.[44]

Tank Training

Postwar reports attested to the U.S. Army's successful creation of both light and heavy tank training programs. Some officers of the AEF Tank Corps, such as Colonel George Patton, promoted the idea of teaching tank crews to cooperate with other arms in combat. A shortage

III. AEF Combined Arms and Open Warfare Training

of vehicles, however, prevented AEF tanks schools from conducting live combined arms exercises. Furthermore, the infantry did not reciprocate by dedicating any of its training to coordinating with tanks on the battlefield.

The AEF's light tank school opened in Bourg, France, in December 1917. American heavy tank crews received instruction in Bovington, England. These schools offered courses in gas engine mechanics, the thirty-seven-millimeter gun, and the Hotchkiss machine gun. Stateside tank training began in January 1918 at Camp Colt, Pennsylvania, under the command of Lieutenant Colonel Dwight D. Eisenhower.

General Pershing requested twelve hundred tanks from the War Department in September 1917. Sadly, only ten American-made tanks made it to France before the armistice. Multiple factors hindered tank production in the United States. The French and British were slow to share their tank specifications with American manufacturers. This left American industry with no design plans for tank engines and armor plate until 1918. The U.S. Army Ordnance Department considered tanks a low priority compared to rifles and artillery and did not assign a central overseer to stateside tank production until late August 1918. Without centralization, five different American companies produced tanks independently and without standard specifications for most of the war. For these reasons, the AEF depended upon the Allies for the majority of its tanks.

The Allies were less than generous when it came to giving tanks to the AEF. The French Army provided the AEF with only 227 light tanks. The British were less charitable with just sixty-four heavy tanks. Needless to say, this tank shortage hampered AEF combined arms training. The American tank school in Bourg did not receive its first ten Renault light tanks until late March 1918.[45]

Despite the obstacles, the light-tank school in Bourg encouraged combined arms by teaching tank crews to familiarize themselves with infantry tactics. Colonel Patton submitted the *Army Tank School Report* on 22 November 1918. He blamed failures in battlefield coordination between tanks and infantry on the infantry's ignorance of tank operation and tactics. Patton asserted: "General training was carried out always with the understanding that tanks exist solely for the purpose of helping the advance of the infantry.... Instructions in carrying out the end sought in this respect were obtained through maneuvers with any infantry obtainable.... Much better liaison with infantry could have been obtained had the infantry ... been given a similar state of preparatory maneuvers with tanks."⁴⁶

The majority of AEF infantrymen were completely uninformed about tanks.

For most American soldiers during the Great War, tanks represented nothing more than armored shields to advance behind. Many AEF officers told the troops under their command that the Germans would flee at the sight of attacking tanks. These claims may have helped boost the infantry's confidence in the "invincibility" of tanks, but were largely untrue. In reality, many American infantrymen soon learned to avoid walking alongside or directly behind tanks, as the vehicles tended to be the preferred targets of German artillery.

Air Support Training

Published doctrine detailing aircraft support for ground forces was scarce during the Great War. Flying low to observe, strafe, and bomb ground targets placed pilots in danger by exposing them to antiaircraft fire. Providing this kind of close air support also diverted attention away from enemy aircraft and resulted in the loss of control over the skies above the battlefield.

Despite the lack of published close air support doctrine during the Great War, the U.S. Army Air Service made progress toward improving coordination with ground forces. The air service designated different aircraft for different tasks and created specialized training programs for each type. General Pershing's wartime orders directed all arms of the AEF to work with the air service. The air service's combined arms efforts were not without flaws. There was a shortage of aircraft for training and unreliable modes of communication. Furthermore, many of the air service's combined arms initiatives began too late to impact the battlefield.

The U.S. Army Air Service Training Section formed in July 1917. The U.S. Army Air Service conducted most of its training at aerodromes

III. AEF Combined Arms and Open Warfare Training

in Issoudun and Gondrecourt, France. Flight training was hazardous, as one out of every eighteen American airmen died during training exercises in France. For this reason, seasoned flight instructors were in high demand. The air service gained experienced leadership when pilots from the Lafayette Flying Corps, which included the famed 103rd "Lafayette" Squadron, joined its ranks as instructors in fall 1917. Captain Frederick W. Zinn, who had served as an American volunteer with the French air service since 1914, was responsible for dispersing the 265 Lafayette Flying Corps personnel throughout the new U.S. Army Air Service. Unlike other arms of the AEF, the U.S. Army Air Service was in no hurry to replace British and French training officers with Americans. The American air service welcomed experienced Allied instructors during the war.

The AEF created bomber, observation, and pursuit aircraft categories and established separate training regimens for each one. It took approximately six months to prepare a trainee for aerial combat. For example, observation aircraft instruction began in the fall of 1917, but the first American observation squadron did not take to the air until April 1918. The bomber school opened in December 1917, but American aerial bombardments did not begin until June 1918. The bomber school did not receive the DeHavilland bombers used at the front until September 1918, so most American bomber crews did not train in the type of aircraft they flew in battle. In the summer of 1918, the air service established flight training facilities in the United States. Few of its graduates made it to France before the end of the war.

The AEF's observation balloon school opened in spring 1918, while stateside balloon training started at Fort Omaha, Nebraska, Fort Monroe, Virginia, Fort Wise, Texas, and Fort Sill, Oklahoma. The U.S. Army used telephones to connect artillery crews with observation balloon spotters during the Spanish-American War. The direct telephone line between artillery batteries and observation balloon crews was the best source of real-time information during the Great War as well. The AEF's seventeen balloon companies used oblong Hydrogen French Caquot Type R kite balloons. These models came with fins for better stability than older spherical balloons. Observation balloons were located between one and two miles behind the front and flew up to four thousand feet in the air. The two-man crew could observer up to forty miles in a fully ascended balloon. Colonel Charles D. Chandler commanded the AEF's Balloon Section. Over twenty observations balloons supported the AEF during the Meuse-Argonne Offensive. Logistical problems plagued Chandler's balloon section. The observation balloon service only had 40 percent of the motor transport it required in November 1918. Without trucks, it was impossible to tow observation balloons to where they were needed along the front.[47]

By 1918, General Pershing issued orders pertaining to the coordination of air power with ground forces. His 6 May 1918 *General Orders Number 70* directed observation balloon trainees to complete fifty hours of instruction with the artillery before entering combat. The commander-in-chief's 29 May 1918 *General Orders Number 81* assigned senior air service officers to all divisions and ordered them to work closely with infantry and artillery commanders in battle planning.[48]

Chief of the Air Service, Major General Mason M. Patrick, produced one of the AEF's most extensive final reports. His report provided evidence of combined arms training in the air service. Aerial support for ground forces, particularly observation, became the focus of air service instruction by September 1918. The air service taught "cavalry reconnaissance," which involved flying at low altitudes, identifying enemy machine-gun positions, and reporting their locations to the infantry. There was also long-range reconnaissance instruction for photographing enemy positions deep behind the lines for division commanders and heavy artillery units. The air service urged all observation and reconnaissance crews to strafe and bomb targets of opportunity during these missions. The air service finally deployed a group of three squadrons dedicated entirely to low altitude strafing during the Meuse-Argonne Offensive.

Like all arms of the AEF, the U.S. Army Air Service combined arms training had its limitations. General Patrick claimed that coordination between the air service and ground forces never reached its full potential during the war. He stated, "Infantry liaison was early attempted and continually used, but even at the close of hostilities it had not been perfected." The majority of aerial observation crews remained unfamiliar with infantry and artillery tactics and operations.

III. AEF Combined Arms and Open Warfare Training

Fifth Division Headquarters at Cunel receives a message dropped from the air. Many observation aircrews communicated with infantry and artillery units on the ground by dropping written messages from the air during the Meuse-Argonne Offensive. Ground troops used colored panels to mark their positions for the pilots. This method of communication was often more reliable than the wireless radios of 1918. Dropping messages from the air continued to be the most common way American airmen communicated with ground forces until the Second World War.

This left them largely ignorant of the capabilities of the arms for which they were providing vital reconnaissance. Furthermore, Patrick's report identified a major gap in air service training concerning the processing of reconnaissance data. The air service handled approximately ten thousand reconnaissance photographs every day during the Meuse-Argonne Offensive. Teams of aerial observers compared new and old photographs to identify changes along enemy lines. Despite the technical nature of this procedure, the air service never opened a photographic analysis school.

The air service experienced communication difficulties as well. The British outfitted their reconnaissance aircraft with spark-gap wireless radio devices in 1917. American reconnaissance aircraft employed the same technology. The war ended, however, before the AEF established a plan to streamline all wireless radio traffic through the Signal

Corps. This posed a major problem for the air service. American wireless radio liaison between the air service and the artillery required relays from air service radio operators to signal corps operators to artillery operators. Not surprisingly, this resulted in countless confused and lost transmissions critical to combined arms coordination. Most aerial observation crews preferred dropping written messages from their aircraft rather than trust this unreliable system.[49]

Engineer Training

General Pershing's wartime orders and the U.S. Army's postwar engineer reports revealed that the majority of American engineer training dealt with supporting the artillery and infantry as a combined arms component. Ultimately, it was the engineers who contributed the most to supporting the AEF's advance during the Meuse-Argonne Offensive. They accomplished this despite the tendency of many division commanders to use their engineers as auxiliary infantrymen during attacks.

The U.S. Army Engineer School opened in October 1917 near Chalons-sur-Marne, France. Its five-week program for officers and noncommissioned officers emphasized combined arms. Trainees underwent three weeks of artillery orientation, including flash and sound ranging, and spent two weeks learning bridge construction. These two combined arms functions made the engineers essential to effective artillery fire and the continuous progress of the infantry.

Pershing's *General Orders Number 77*, dated 24 May 1918, instructed the U.S. Army Engineer School to place more emphasis on teaching trainees to support the infantry by repairing roads and bridges, rather than building them, and removing battlefield obstacles. The commander-in-chief based this directive on the muddy roads and swollen rivers American troops encountered thus far on the western front. On 7 August 1918, Pershing's *General Orders Number 131* placed sound- and flash-ranging engineers in every American artillery battery.[50] This resulted in improved artillery support for the infantry in time for both the Saint Mihiel and Meuse-Argonne offensives.

Although supporting the infantry and artillery were the primary combined arms functions of American engineers in the Great War,

III. AEF Combined Arms and Open Warfare Training

their crowning achievement was in transportation. In April 1919, Colonel G. A. Youngberg quantified the AEF's rail systems, locomotives, rolling stock, roadways, and fueling stations in his *Engineer Report*. U.S. Army engineers built 947 miles of standard railroad tracks, maintained nearly fourteen hundred miles of light railroad tracks, and operated 165 locomotives and 1,695 rail cars by 11 November. Standard rail lines transported American troops, ammunition, and supplies to the combat zone, while light rail lines moved ammunition and supplies within the combat zone. American engineers maintained an additional thousand miles of light tracks captured from the Germans. Furthermore, the engineers built ninety roads, repaired three hundred roads, and operated seventeen fueling stations during the conflict. General Pershing's Assistant Chief of Staff of the Services of Supply, Brigadier General George Van Horne Moseley, praised the railroad work of AEF engineers after the war. He claimed that the engineers "rendered very satisfactory service in taking part of the burden off other means of transportation."[51]

Motor Transport Capabilities

The U.S. Army purchased its first truck in 1907 and owned only a dozen by 1911. The army found these automobiles ideal for moving troops, supplies, and wounded behind the lines. General Pershing used about four thousand Dodge trucks to carry ten thousand tons of supplies over a four hundred-mile distance during the Punitive Expedition into Mexico. The U.S. Army's use of motor transport in Mexico hardly compared to the tens of thousands of vehicles that transported American soldiers and material in France in 1918. The AEF's 1917 organization plan called for all combat and support elements to be fully motorized, requiring about fifty thousand vehicles. The United States had already manufactured and sold forty thousand automobiles to Great Britain and France since 1914. Many of Pershing's military and political contemporaries questioned the economic feasibility of his request for fifty thousand more vehicles. Needless to say, full motorization never materialized for the U.S. Army during the Great War.

The Inspector General of the U.S. Army Motor Transport Corps,

American Tactical Advancement in World War I

Colonel C. C. Carson, commented on AEF motorization in his final report in April 1919. The U.S. Army received motor vehicles from 294 different manufacturers. Ford, Dodge, Cadillac, General Motors, Packard, and White were the AEF's most common makes. Harley-Davidson, Excelsior, and Cleveland supplied the U.S. Army with its motorcycles. In November 1918, the AEF possessed eighty-one cars, ninety-six trucks, twelve wheeled tractors, six caterpillar tractors, thirty-seven trailers, ten fuel tankers, three reconnaissance vehicles, and sixteen motorcycles.[52] These numbers fell far short of General Pershing's full motorization goal.

General Moseley's final Service of Supply report in April 1920 noted the strengths and weaknesses of AEF transportation at the end of the war. Unlike the American rail systems in France, Moseley judged the AEF's motor transport as insufficient. He reported, "At the signing of the Armistice ... we were short of about half our motor vehicles." Furthermore, the chronic spare-parts shortage resulted in consistently poor inspection records for the AEF's motor transport corps.[53] Motor transport deficiency represented a serious shortcoming in the AEF's ability to support and sustain its combined arms operations during the Meuse-Argonne Offensive.

Communication Capabilities

The future of the U.S. Army Signal Corps was in question after the American Civil War. Its only post–Civil War responsibilities were semaphore and reporting the weather. The army even considered folding the Signal Corps into the engineers. The early 1900s, however, brought the Signal Corps a new set of tasks when the army adopted the telegraph and telephone for field use. The U.S. Army's first dedicated signal school opened in 1905 at Fort Leavenworth. The Signal Corps managed all AEF rear-echelon communications and maintained communication equipment for all frontline infantry and artillery units during the Great War.

Communication was the only aspect of the AEF where capability exceeded need. Postwar reports illustrated the AEF Signal Corps' vast surplus of equipment. American communications did face some

III. AEF Combined Arms and Open Warfare Training

technological constraints, however, as the electronic communication devices of 1918 were not secure and virtually immobile. Chief Signal Officer, Major General George O. Squier, oversaw Western Electric's development and testing of a portable wireless telephone in Chicago, Illinois in 1918, but the device was not ready for the battlefield until after the war ended. Technological limitations caused many commanders to utilize a variety of communication forms, such as heliographs, signal lamps, runners, and pigeons. This lack of standardization led to frequent communication breakdowns and hampered AEF combined arms. For example, there was no means for frontline infantry units to call for instantaneous adjustments of artillery fire, tank crews relied upon pigeons to carry messages back to headquarters, and commanders issued orders without up-to-the-minute reports from the combat zone.

The *Signal Report* of March 1919 revealed the AEF's extensive communication capabilities. Each AEF division contained one signal battalion. The Signal Corps ran sixty thousand miles of wire, maintained sixteen thousand telephones, and operated eight thousand switchboards during the war. Women auxiliaries worked most Signal Corps switchboards behind the lines. Over three hundred of these "Hello Girls" served in France during the war. The U.S. Army successfully connected its division, corps, and general headquarters with spark-gap wireless telegraph. With their large antenna and limited six-mile range, these wireless units were only useful in rear areas. The AEF employed the 1915 Fullerphone wired telegraph device along the front. Spark-gap did not allow for simultaneous signals to travel along one wire without causing interference. Therefore, at least two wires were needed to establish connections between two points. Telegraph operators, depending upon their skill level, could transmit between fifteen and forty words per minute in Morse Code. According to the report, the Signal Corps had enough surplus equipment to provide communications for an additional forty divisions when the war ended. Inspector General Brewster credited the Signal Corps with the best wartime inspection record of the entire AEF.[54]

No form of communication was completely reliable or secure against enemy monitoring, especially in the combat zone. Enemy artillery fire frequently severed telegraph and telephone wires. U.S. Army regulations recommended communications wires be buried at least six feet underground to protect them from enemy artillery fire. In reality, frontline commanders rarely ordered their troops to spend the time necessary to bury wires six feet deep. As a result, damaged communication wires were common. Furthermore, wireless devices were dodgy at best. Because of the use of ground returns, it was possible for the enemy to listen in on wired communication without even splicing into the wires. The Thirty-Sixth Division AEF found a way around this problem by using Choctaw Native Americans as "code talkers." Choctaw was completely undecipherable by German eavesdroppers.

The biggest problem with electronic communication was its lack of mobility. The telegraph, telephone, and wireless radio units of the Great War were cumbersome and fragile. Most had to be moved by horse and wagon. Their battery packs alone usually required their own horse-drawn cart. For this reason, communication in the combat zone combat depended on runners carrying written messages. To insure delivery in heavy combat, commanders often sent up to six runners to deliver a single message. The AEF made wide use of pigeons as well. Messenger pigeons bore identification numbers on their wings and carried written notes attached to their legs from units in the combat zone to lofts behind the lines. Pigeons were capable of "remembering" the location of their loft for up to four days. The U.S. Army maintained a pigeon training center at Fort Monmouth, New Jersey, during the war. Needless to say, the AEF's overall lack of a standard mode of communication at the front made it difficult to collect, collate, and distribute tactical data. In spring 1918, officers of the Twenty-Seventh Division took crash courses in the use of telephones, telegraph, wireless radio, message runners, and messenger pigeons to prepare them for the numerous modes of communication they would encounter on the battlefield.[55] The AEF gathered massive amounts of information through all these various methods, but never developed a method for synthesizing it. The real time communication technology best suited to combined arms warfare simply did not exist in 1918.

III. AEF Combined Arms and Open Warfare Training

Open Warfare Training

U.S. Army tactical training eventually incorporated some principles of General Pershing's open warfare concept, such as infiltration, fire-and-maneuver, use of cover, and battlefield initiative. By spring 1918, the Seventy-Ninth Division's instruction included "capturing strategic points," "outmaneuvering strong columns," and "surprising unsuspecting encampments."[56] Unfortunately, the antiquated concept of fighting spirit and an overemphasis on marksmanship lingered in the AEF's training regimen as well. For example, a bulletin from General O'Ryan to the Twenty-Seventh Division on 9 June 1918 contained long sections on the importance of *élan* and marksmanship in the battles ahead.[57] As in the case of combined arms, the end of the war ultimately interrupted the AEF's formal education in open warfare tactics.

U.S. Army training started featuring aspects of Pershing's open warfare concept before he clarified his version of the tactics in September 1918. In September 1917, General Alvord established fire-and-maneuver as a major component of AEF instruction in France. In December 1917, Pershing cabled Alvord to urge that open warfare be the focus of stateside instruction as well. At the same time, they both ordered marksmanship to supersede all other forms of AEF training. Most American conversations regarding infantry tactics were rife with these kinds of paradoxes, as the key principles of nineteenth century combat slowly gave way to new techniques. For example, General Pershing's October 1917 cable to Alvord affirmed that rifle proficiency was central to successful open warfare tactics. In addition to marksmanship, the commander-in-chief's *communiqué* established *élan* as a major tenet of open warfare training.[58]

The *Instructions for the Training of Platoons for Offensive Action, 1917* stressed uniformity in tactical doctrine, training, and battlefield conduct. The directive called for standardized training in the use of cover for all AEF platoons and urged platoon leaders to learn to seize the initiative in battle without waiting for orders from above. The AEF's infantry specialist school also incorporated fire-and-maneuver and battlefield initiative and improvisation as standard learning goals by 1918.[59] The Twenty-Seventh Division practiced fire-and-maneuver while preparing for the Meuse-Argonne Offensive in September 1918. General O'Ryan described "enemy machine-gun posts ... indicated by small red flags ... were reduced by fire action from the front, while ... enveloped or attacked from the rear."[60] These efforts to standardize training across the AEF were reflections of the Uptonian mindset from the pre-war era.

General Liggett recalled open warfare as an objective of AEF training in his postwar reflections. He endorsed an Upton's philosophy concerning battlefield survival through effective leadership. Liggett referred to this concept as "intelligent discipline" and promoted it in officer and noncommissioned officer training throughout First Army. "Intelligent discipline" was the balance between knowing when to follow orders and when to take the initiative in battle. Liggett did not, however, tolerate recklessness. He warned his officers to be mindful of casualty rates and wanted them to understand that dead and wounded soldiers had no value on the battlefield.[61] Following the Uptonian tradition of the prewar period, AEF officer and noncommissioned officer instruction promoted aggressive leadership as the key to effective battlefield performance. Hasty training, however, prevented most AEF officers from developing into the independent battlefield tacticians Upton envisioned. In fall 1918, General O'Ryan observed how many of the Twenty-Seventh Division's company and platoon commanders still needed to take the initiative more often and seize opportunities to gain additional ground rather than wait for directions from their superiors.[62]

Conclusion

The AEF clarified its own tactical training regimen by contrasting it with French and British instruction. American instructors criticized the Allies for being unimaginative and mired in stagnant trench warfare. Meanwhile, General Pershing gradually devised his own open warfare tactics and worked to standardize its training across the AEF. He integrated his open warfare concepts of fire-and-maneuver, the use of cover, and battlefield initiative into American instruction. The commander-in-chief also promoted combined arms by encouraging the

III. AEF Combined Arms and Open Warfare Training

infantry to be proficient in a variety of weapons and urging all arms to coordinate their training and battle planning.

These efforts did not go unhindered. Tanks, aircraft, and artillery were seldom available for instructional purposes. There was a motor transport shortage and inefficient communication across the AEF. Different arms of the AEF integrated new tactics into their training programs at different times, some far later than others. General Pershing's decision to rush soldiers to the frontlines resulted in hasty training for all but one AEF division. Finally, the war ended before many American open warfare and combined arms training initiatives could produce results on the battlefield. These limitations, however, did not completely nullify the benefits of the U.S. Army's advanced tactical instruction. Lieutenant Colonel George C. Marshall believed that the tactical strengths and weaknesses of the AEF by the final phase of the Meuse-Argonne Offensive were similar to those exhibited by any army on the western front at that time. Colonel Patton echoed this sentiment when he stated that the French and British were just as guilty of sending undertrained troops into battle in 1918 as the AEF.[63]

IV

AEF Combined Arms and Open Warfare in Action

Formal training failed to fully prepare the AEF for the tactical challenges of the Great War. AEF divisions, therefore, adapted to modern warfare primarily through survival instinct and combat experience. The final phase of the Meuse-Argonne Offensive from 1 November 1918 until the armistice showcased the combined arms and open warfare tactics the AEF learned since arriving on the western front six months earlier. The performance of the AEF's First Army and its Fifth Division during the last push of the war reflected this learning curve on a tactical level.

The orders, memoranda, reports, and postwar reflections of General Pershing, General Liggett, and Colonel Marshall revealed that the strategic plan for the final phase of the Meuse-Argonne Offensive called for combined arms. These sources also illustrated how American combat units carried out these tactics with moderate success. Coordination between the artillery and infantry figured prominently in Pershing's expectations for what turned out to be the final phase of the Meuse-Argonne Offensive. The AEF prefaced the assault on 1 November with its largest preliminary bombardment of the war. The artillery successfully supported the infantry attack with a creeping barrage for the initial

IV. AEF Combined Arms and Open Warfare in Action

twenty-four hours of the operation. Pershing's battle plan recommended extensive use of chemical agents. The AEF's actual employment of poison gas fell short of the commander-in-chief's hopes, as most division commanders still feared retaliatory chemical attacks. Armor support for the last push also failed to meet Pershing's expectations due to a tank shortage. Air support consisted primarily of reconnaissance, but in some instances, American aircraft bombed and strafed enemy lines as the infantry moved forward. Finally, the infantry's progress relied heavily on engineers, who enabled the AEF divisions to pressure the Imperial German Army right up to the armistice by building roads, railroads, and bridges.

Wartime and postwar sources contained less evidence of the AEF's use of open warfare during the final phase of the Meuse-Argonne Offensive. This was mainly due to the fact that the Imperial German Army conducted an effective fighting retreat from 1 November until the armistice. Confronted with a massive rearguard action, American troops encountered few opportunities for fire-and-maneuver and infiltration. Fire-and-maneuver and infiltration were best suited for attacks against fixed defensive positions and persistent strong points, not pockets of resistance relocating from one day to the next as the Germans fell back during the first eleven days of November. Furthermore, Supreme Allied Commander *Generalissimo* Ferdinand Foch responded to the collapse of the German line by contradicting General Pershing's open warfare concept with orders for all commanders to advance without regard for flanks or local objectives. Many AEF officers eagerly carried out this order in an effort to gain as much ground and glory as possible before the war's end. Needless to say, this reckless disregard for open warfare produced heavy American casualties in areas where German defenses were still resilient.

First Army's Combined Arms Planning and Preparation

General Pershing's *General Orders Number 23-A* placed General Liggett in command of the AEF's First Army on 12 October 1918. In addition to reinforcing and resupplying First Army's depleted ranks,

American Tactical Advancement in World War I

Liggett immediately addressed combined arms. He issued memoranda on 16 and 20 October encouraging his corps commanders to employ chemical agents in the coming assault. Liggett ordered that infantry commanders direct the use of poison gas in battle. This decision ran contrary to the mainstream practice of placing artillery commanders in charge of deploying chemical agents. Liggett believed infantry officers were better informed as to when and where chemical support was needed in combat. His 16 October directive stated that "gas will be employed in the zone of action ... only upon agreement of the commanding general of the corps concerned." Liggett's 20 October follow up gave similar instructions: "Yperite[mustard gas] bombardments will be carefully regulated by corps commanders."

Meanwhile, First Army conducted patrols and raids to reconnoiter enemy defenses and secure an ideal jump-off point for the next phase of the Meuse-Argonne Offensive. Major General John Hines' III Corps captured Bois des Rappes and General Summerall's V Corps took Bois de Bantheville. These successes cleared German machine-gun positions from the wooded terrain in front of First Army by late October. The artillery supported both operations with effective counter-battery fire and mustard gas.[1]

In mid–October, the commander-in-chief ordered the AEF's air forces to make reconnaissance and close air support for First Army their top priority. He spoke with U.S. Navy Rear Admiral Charles P. Plunkett, the commander of the AEF's naval railway batteries, to arrange for fourteen-inch railway-mounted naval guns to be brought in position to support Liggett. General Pershing requested additional horses from the French Army to help keep First Army's artillery advancing with the infantry. The French denied this request due to their own shortage of draft animals.[2]

General Liggett's *Field Orders Number 88* and his extensive *Battle Instructions of October 22, 1918*, detailed First Army's combined arms plans for the final phase of the Meuse-Argonne Offensive. *Field Orders Number 88* called for artillery to concentrate on counter-battery fire at the outset of the operation. Liggett intended this bombardment to suppress enemy artillery opposite the infantry in the wooded heights near Dun, Bois de Sassey, Bois de Barricourt, and Bois de Bourgogne.

IV. AEF Combined Arms and Open Warfare in Action

Thick enemy barbed wire on the approach to Hill 252 near Bantheville. Uncoordinated combined arms and heavy casualties suffered in this region from 14 to 17 October 1918 led General Perishing to replace Major General John E. McMahon with Major General Hanson Ely as Commander of the Fifth Division.

He also directed the artillery to impede the enemy's ability to reinforce by saturating German supply lines with mustard gas. Finally, Liggett ordered his artillery to coordinate a creeping barrage to protect with the infantry and maintain close contact with First Army's corps commanders throughout the entire operation. He stated, "In order to coordinate the fire of the Army artillery with the infantry advance, commanders of Army artillery groups will maintain close liaison with the corps commanders interested."[3]

General Liggett's 22 October battle instructions represented the clearest illustration of his understanding of combined arms by fall 1918 and his effort to incorporate these tactics into his plan for the last phase of the Meuse-Argonne Offensive. *Battle Instructions of October 22, 1918, Annex Number 1: Employment of Army Artillery* provided additional details concerning artillery support for the infantry. In this document, Liggett identified decoy artillery targets intended to confuse enemy defenders during the opening assault. The annex allotted for a two-hour preliminary bombardment, the use of mustard gas to contain German troop and supply movements, and a smoke screen to obscure American attackers from enemy artillery and machine guns.[4]

Battle Instructions of October 22, 1918, Annex Number 2: Plan for Employment of Air Service Units, American First Army revealed General Liggett's expectations for air support in the coming attack. His airpower annex called for both day and night bombers to concentrate on German aerodromes and rail lines around the towns of Mézières, Buzancy, Dun, Sedan, Stenay, and Baumont prior to the operation. This would help insure American air superiority and weaken German logistics by the start of the offensive. Meanwhile, Liggett ordered continuous reconnaissance sorties so that the infantry and artillery would have extensive photographs and maps of German defenses. His plan instructed pursuit aircraft to fly over enemy territory just ahead of the infantry assault to eliminate German reconnaissance aircraft, destroy observation balloons, and harass enemy troops, supply depots, and machine-gun positions. Liggett clearly stated, "The general role of the Air Service will be ... active close cooperation with the infantry on the battlefield."[5]

General Liggett's *Battle Instructions of October 22, 1918, Annex Number 3: Plan of Employment of Special Gas Troops* attached soldiers from the First Gas Regiment to each division in First Army to assist with the employment of smoke and poison gas during the offensive. Liggett wrote, "These troops will assist the divisions with which they are operating, both in the preparation for and the progress of the attack." *Battle Instructions of October 22, 1918, Annex Number 5: Plan of Employment of Engineer Troops, Supply of Engineer Material, and Water Service* ordered engineers to construct light rail lines immediately behind the advancing infantry and utilize as much captured material as possible to keep the infantry moving forward and supply lines open during the attack.[6]

Liggett's *Battle Instructions of October 22, 1918, Annex Number 7: Means of Information* made all corps signalmen responsible for telegraph and telephone communication in their sectors during the assault. General Liggett recognized that maintaining communication between front line units and rear command centers would be a major challenge as the attack gained momentum. He urged advancing units to use wire

IV. AEF Combined Arms and Open Warfare in Action

communication for as long as possible, even if it meant scavenging German equipment, and recommended that units beyond wire contact resort to semaphore and carrier pigeons.[7]

First Army's Combined Arms in Action, 1–11 November 1918

General Pershing, General Liggett, and Colonel Marshall's orders, memoranda, reports, and postwar reflections illustrated their appreciation for combined arms by the final phase of the Meuse-Argonne Offensive. These sources also identified the obstacles limiting the AEF's ability to conduct combined arms during its last attack of the Great War. Artillery support for the infantry was the most common form of combined arms practiced by First Army from 1 November until the armistice. Unfortunately, manpower shortages forced generals Pershing and Liggett to amalgamate understrength artillery and infantry units just before the operation. Many artillery crews found themselves paired with infantry units whom they had neither previously trained nor fought together. This compromised the familiarity, trust, and cohesion necessary for effective battlefield coordination between foot soldiers and artillerymen. Furthermore, the speed of First Army's advance during the final phase of the Meuse-Argonne Offensive caused communications to lag and supply lines to break down. A shortage of both horses and motor transport compounded these problems. By 11 November, the AEF's entire supply train stretched over five hundred miles from Bordeaux to the Franco-German border.

General Pershing set 1 November as the launch date for the Final Phase of the Meuse-Argonne Offensive on 26 October. First Army used the following five days for last-minute preparations, such as stockpiling ammunition, constructing additional rail lines, and connecting extra communication wires. During this time, General Liggett ordered his seventy-five-millimeter cannons and thirty-seven-millimeter guns forward to support the infantry with direct suppressing fire on enemy machine-gun positions during the attack's jump-off.[8]

The days preceding the initial assault were not without complications for First Army. On 25 October, *Generalissimo* Foch issued an order negating the employment of the limited-objective aspect of General Pershing's open warfare concept during the last push. *Foch's Views on Limited Objectives*, instructed General Liggett's First Army to "advance as far as possible ... without any attempt at alignment." According to the decree, "limiting beforehand the progress of troops to lines placed on the map ... compromises the final result."[9] Foch's last minute order, coupled with the Imperial German Army's dogged fighting withdrawal, limited the AEF's use of Pershing's open warfare tactics during the final phase of the Meuse-Argonne Offensive. Furthermore, consistently poor weather prevented the air service from conducting reconnaissance sorties during First Army's last days of planning. This left Liggett without the most recent maps, photographs, and intelligence reports while making final adjustments for 1 November.

Despite the obstacles, General Liggett was able to perform some combined arms preparations. The weather improved enough to allow First Army's aircraft to complete some bomber and reconnaissance sorties before 1 November. For example, Colonel Marshall's *Operations Report Number 25* claimed that American bombers damaged German defenses around Briquenay on 27 October. His *Operations Report Number 27* reported another American aerial bombing raid on 29 October. According to Marshall, observation aircraft took five hundred reconnaissance photographs that day as well.[10] General Pershing's Assistant Chief of Staff of Intelligence, Brigadier General Dennis E. Nolan, claimed that American bombers in support of First Army dropped their ordnance on enemy roads and supply dumps around the towns of Tailly, Barricourt, and Villers-devant-Dun on 30 October. These aircraft also strafed German vehicle convoys and troop concentrations along the way. Marshall described an additional five and a half tons of bombs dropped and 450 reconnaissance photographs taken on 31 October in *Operations Report Number 28*. Liggett conducted these air operations despite major aircraft and spare parts shortages. First Army had less than eight hundred aircraft at its disposal on 1 November, about half of what it had less than two months earlier.[11]

The final phase of the Meuse-Argonne Offensive began in the early hours of 1 November. General Nolan called it "the perfect cooperation of all arms—infantry, artillery, airplanes and tanks."[12] A two-

IV. AEF Combined Arms and Open Warfare in Action

hour preliminary bombardment opened around 3:30 a.m. General Liggett had 2,466 artillery pieces, including mortars, cannons, and howitzers, at his disposal that morning. The intelligence gathered by observation aircraft during the preceding days helped the artillery focus its preliminary bombardment on German artillery batteries, ammunition depots, crossroads, railway junctions, and strong points near Bois de Bourgogne. First Army's preliminary bombardment also included chemical agents and Rear Admiral Plunkett's naval railway guns. Following what was standard procedure by this point in the war, the artillery provided the infantry with a smoke screen and an effective creeping barrage when they began their advance north toward the Meuse River around 5:30 a.m. Colonel Lanza, First Army's Chief of Artillery Operations, described the close artillery support as "a rolling barrage ... which scoured the terrain long in advance of the first infantry lines."[13]

Airpower and armor also contributed to Liggett's 1 November combined arms operation. Chief of First Army Air Service, Brigadier General William Mitchell, set close air support for the advancing infantry as the priority that day. American aircraft drove most German pursuit planes out of the sector prior to the attack and commanded the skies above the battlefield. This air superiority allowed American aircraft to fly reconnaissance and bomb and strafe German machine-gun and artillery positions without discretion as the infantry advanced.[14] General Liggett's *Special Orders Number 518* attached the Third Tank Brigade to General Summerall's V Corps for the opening attack on 1 November. Colonel Marshall's *Operations Report Number 29* described how fifteen of these tanks helped Major General John A. Legeune's Second Division break through the German lines at Cléry-le-Grand and Aincreville by the end of the day.[15]

First Army's 1 November combined arms assault incorporated some of General Pershing's open warfare tactics as well. With the artillery effectively suppressing enemy strong points, the majority of Liggett's attackers were able to bypass these positions. The U.S. Army War College's *Report of First Army* stated that "the major portion of the divisions pushed rapidly on between these strong points. This scheme worked admirably."[16]

American Tactical Advancement in World War I

On 2 November, German troops opposing Major General Joseph T. Dickman's I Corps at Champigneulle fell back so rapidly that his infantry needed to use trucks to maintain contact with the enemy. Unfortunately, these swift territorial gains resulted in communication breakdowns. First Army Headquarters at Souilly lost contact with many frontline divisions as early as 3 November. Despite the use of runners, pigeons, motorcycle couriers, and messages dropped from aircraft, it took between four and five hours for General Liggett to receive updates from his commanders in the combat zone. Signalmen worked in vain to extend and repair the wires connecting Liggett with his division commanders. Colonel Marshall's 4 November *Operations Report Number 32* described the situation: "Due to haste in which the enemy retired and the difficulty in communication with our advancing troops, it was impossible to outline even approximately the hostile front line at a certain hour."[17]

Deteriorating communications did not completely prohibit General Liggett from carrying on with his combined arms effort after 1 November. First Army's artillery continued to support its infantry divisions by conducting indirect fire on strategic targets behind enemy lines. This interfered with the Imperial German Army's ability to organize an effective defense against First Army's offensive. General Nolan's communiqués described heavy artillery bombarding German railway junctions in the towns of Montmédy, Longuyon, and Conflans on 3 November. Liggett's *Field Orders Number 102* directed heavy artillery to concentrate fire on the German rail lines from Montmédy to Mouzon in support of III Corps' attack on Côte Saint Germain and Murvaux on 6 November. In his *Operations Report Number 35*, Colonel Marshall noted how artillery saturated the heights of Borne du Cornouiller with chemical agents on 6 November to harass German machine guns and artillery positioned on this elevated terrain overlooking First Army's primary lane of attack.[18]

First Army continued to receive similar support from the air. General Nolan's communiqués contained numerous accounts of American aircraft bombing strategic assets behind enemy lines. Bomber aircraft struck German roads and rail lines near the towns of Montmédy, Mouzon, and Raucourt on both 4 and 5 November. Pursuit aircraft

IV. AEF Combined Arms and Open Warfare in Action

strafed German troop concentrations around these towns as well. According to Nolan, American aerial bombardments on 6 November destroyed large enemy ammunition depots at Bealon, Remoiville, and Gibercy.[19]

By 6 November, First Army's logistical and communication problems became critical. On 3 November at Dun, General Hines' III Corps was the first element of Liggett's army to cross the Meuse River. General Nolan noted how First Army's engineers struggled from then onward to build and maintain foot and pontoon bridges across the Meuse River under constant German artillery and machine-gun fire from the east bank of the river. Marshall's war memoirs recalled First Army machine-gun crews pulling their weapons across the battlefield with improvised handcarts due to a shortage of horses after 5 November. Communication breakdowns between divisions' headquarters and their advanced units resulted in friendly artillery falling on AEF troops around the town of Brandeville on 6 November.[20]

Meanwhile, General Pershing anticipated the coming armistice. In keeping with *Generalissimo* Foch's directive, the commander-in-chief decided to apply maximum pressure on the Germans to achieve more favorable terms for the Allies. He believed an invasion of Imperial Germany would force an unconditional surrender. On 5 November, Pershing ordered General Liggett to advance First Army at all costs without regard for flanks or local objectives in an effort to defeat all German resistance west of the Meuse River. This sense of urgency exacerbated Liggett's already precarious supply, logistical, and communication situation. Colonel Marshall's *Operations Report Number 34* observed First Army's strained supply lines on 5 November: "Unusually rapid progress, involving large gains, was made along the entire front west of the Meuse and a deep advance was carried out by the III Corps east of the Meuse.... Supply of troops of I and V Corps is difficult." Marshall reported the supply situation becoming even more "difficult" over the next four days.[21]

News of the imminent armistice reached First Army Headquarters in Souilly at 6:00 a.m. on 11 November. Despite Marshall's best efforts to notify all AEF divisions of the ceasefire, several frontline units did not receive the news until after the 11:00 a.m. ceasefire. This meant the communications lag from headquarters to the furthest reaches of the combat zone exceeded five hours on the day the war ended. The armistice could not have come at a more fortuitous time in respect to General Liggett's precarious communication and logistical predicament. From Colonel Lanza's perspective, the horses and motor transport for First Army's artillery "lasted until the Armistice, but it was doubtful if they would have lasted much longer."[22] Supply, logistical, and communications difficulties not only hampered First Army's ability to execute combined arms, but threatened the AEF's entire campaign. First Army would require a pause to rest, regroup, and refit if it was to continue its push toward the Franco-German border after 11 November. A break such as this, may have allotted the Imperial German Army enough time to establish a formidable defensive line along the Rhine River and drag the war into 1919.

First Army's German Opposition

Four years of attrition had worn down the Imperial German Army by fall 1918. Only twelve German divisions on the western front were at full strength in November 1918. German units facing First Army during the final phase of the Meuse-Argonne Offensive found themselves outnumbered three-to-one and incapable of counteroffensive action. Chief of the Imperial German General Staff, Field Marshal Paul von Hindenburg, and German Army Deputy Chief of Staff, General Ludendorff, had abandoned the notion of winning the war and planned for a piecemeal fighting retreat in August 1918. General Ludendorff issued an order restricting his armies to defensive maneuvers and initiating a rearguard action toward the Franco-German border on 30 September.

Despite its state of desperation, the Imperial German Army's defensive strategy was sound and conducted with enough efficiency and vigor to challenge First Army during its last push of the war. German division commanders shortened their lines, defended key railroad junctions, such as Sedan, and conducted a determined fighting retreat toward Germany. As General Ludendorff recalled after the war: "The defensive battle on the Meuse had followed a favorable course, in spite

IV. AEF Combined Arms and Open Warfare in Action

of the absolutely overwhelming superiority of the enemy. The enemy gained ground, but slowly."[23] The Imperial German Army perfected a defense in depth technique by 1917. This tactic called for defenders to allow attackers to capture lightly-manned forward positions so they could decimate them with dense machine-gun and artillery fire from areas further in the rear. General Liggett's army faced this type of defense in depth along the Meuse River in November 1918.

The Imperial German Army also enjoyed a geographic advantage along the Meuse River. The German forces opposing First Army occupied a favorable defensive position on the elevated forest terrain overlooking the river. Ludendorff recollected, "General Headquarters had to reckon with the possibility of withdrawing the front back to the Meuse line at the beginning of November in order to still further shorten it." Generals Hindenburg and Ludendorff hoped this defensive strategy would hold off the Allies long enough to force favorable peace terms. Ludendorff called up an additional 600,000 troops to serve on the western front in late October and directed all army groups to make a strategic retreat to the Meuse River, referred to as the *kriegsmarch*, on 4 November.[24] General Liggett's forces partially nullified Hindenburg and Ludendorff's plan by crossing the Meuse that same day, but this did not completely curtail German resistance on the east bank of the river.

The AEF's impression of the enemy forces they faced in November 1918 was more significant than the actual condition of the Imperial German Army at that time, as it was this perception that drove First Army's tactical planning. Well-placed machine guns and pre-registered artillery compensated for the German Army's lack of manpower and convinced American commanders not to underestimate their adversary. General Pershing was aware that the German divisions facing the AEF in November lacked reserves, but he believed their experience and favorable defensive position made up for this deficiency. His respect for the enemy trickled down to his commanders. After the war, General Liggett commended General Georg von der Marwitz, the commander of the German Fifth Army, for orchestrating an effective fighting retreat without the benefit of motor transport and having only twelve combat-ready divisions under his command. Marwitz issued a general order to his army on 1 November instructing all commanders to utilize the cover of night to retreat from towns and villages, destroy all bridges across the Meuse, and use artillery to slow the pursuing Americans.[25] The entire German Fifth Army crossed the Meuse by 3 November. Colonel Huidekoper of the Thirty-Third Division recalled as late as 10 November that "prisoners did not come forward and give themselves up. Machine gunners were very active in occupying flanking positions and withdrawing promptly when threatened by envelopment."[26]

The skillful German withdrawal to the east bank of the Meuse masked the crisis unfolding behind their lines, where morale decreased and rebellious units refused to fight. Most German soldiers lost their will to attack by the fall of 1918, although many were still prepared to defend against an invasion of their homeland. In this respect, the Imperial German Army of late 1918 was similar to the mutinous French Army of spring 1917. General Ludendorff's postwar memoirs admitted, "The number of shirkers behind the front increased alarmingly. The men who fought in the front line were heroes." The AEF was not convinced of the inevitability of Imperial Germany's collapse in November 1918. For example, the Thirty-Third Division received a bulletin regarding opposing troops on 8 November stating that interrogated German prisoners "have not heard any project of retreat mentioned." A German officer captured by the Thirty-Third on 10 November proclaimed that the Imperial German Army "can and will hold this line unless the spirit manifested by the troops in the past four years is broken," something he swore was impossible.[27] The bold claims of this young German officer rang true, as stubborn German machine-gun and artillery fire drove the Thirty-Third Division back at Saint Hilaire, Wadonville, Bois de Warville, Riaville, and Fresnes on 8 and 9 November.[28]

Colonel Marshall suspected that the armistice negotiations were a German ploy to buy time to form a new defensive line along the Rhine River and prolong the war into 1919. This suspicion was not totally unfounded, as Hindenburg, Ludendorff, and German Chancellor Prince Max von Baden had considered such a plan in late October. The scenario involved a "scorched earth" retreat across the eastern portions of France and Belgium and the use of the German border fortresses

IV. AEF Combined Arms and Open Warfare in Action

around Metz to stall the American invasion. The Imperial German War Ministry even suggested that civilians prepare to resist the foreign invaders with *guerrilla* warfare. These conversations ultimately failed to convince the government and military that continued resistance would achieve a more favorable outcome for Germany. Despite this desperation behind the German lines, Marshall observed in his *Operations Report Number 38* on 10 November, the day that Kaiser Wilhelm II abdicated, that opposing German artillery and machine-gun fire seemed to be getting more intense and better organized.[29] First Army experienced the final phase of the Meuse-Argonne Offensive as an effective German rear guard battle, not a reckless general retreat.

Fifth Division's Combined Arms Planning and Preparation

The War Department ordered the formation of the Fifth "Red Diamond" Division in November 1917. Citizen soldiers, volunteers from the eastern United States, arrived at Camp Logan all through the winter to bring this Regular Army division up to combat strength. Camp Logan was one of many temporary training facilities hastily constructed by the U.S. Army after the declaration of war in April 1917. Weapon shortages limited the Fifth's training at Camp Logan. For example, fist-sized concrete balls served as grenades, sawhorses substituted for machine guns, and tent pegs represented artillery positions. French and British training officers eventually arrived to teach the men of the Fifth Division how to operate gasmasks, occupy trenches, and fire rifles. Major General John E. McMahon took command of the division in January 1918. McMahon rose through the ranks as an artillery officer.[30] His appointment to lead an infantry division was symptomatic of the U.S. Army's shortage of senior officers during the Great War.

The Fifth Division sent officers and noncommissioned officers from each of the division's two infantry brigades, the Ninth and the Tenth, to the U.S. Army Infantry Specialist School at Chatillon-sur-Seine, France in February 1918 as an advanced training detachment. General Pershing and the War Department expected these seventy-seven men to train the rest of the division when it reached Europe.

American Tactical Advancement in World War I

Four hundred officers and noncommissioned officers form the division's artillery brigade, the Fifth, went to the AEF's field artillery school at LeValdahon, France for advanced training as well. This was their introduction to the 155-millimeter howitzers and 75-millimeter cannons they would use in combat on the western front. The Fifth Division's engineer regiment, the Seventh, conducted its training at Fort Leavenworth. A letter from Lawrence L. Arbuckle, a private in the Seventh Engineers, testified to their inadequate training. Arbuckle wrote that he had finally fired on a rifle range for the first time on 18 February, less than one month before departing for Europe.[31] In March, the Seventh Engineers Regiment was the first element of the Fifth Division to land in France. The rest of the division arrived piecemeal until June, when it became the eighth AEF division to assemble in Europe.

Although the Fifth Division spent the prescribed six months in a stateside training facility, its piecemeal assembly at Fort Logan and fragmented shipment to Europe led to discrepancies in the amount and type of instruction received across the division. First Lieutenant Vernon G. Olsmith took command of D Company of the First Battalion of the Sixth Infantry Regiment less than two weeks before it left for France. His postwar recollections noted the short time he had to train and get acquainted with his new unit as "a hectic ten days, as we rushed to complete certain pre-embarkation training." In his memoirs, he described the poor quality and inexperience of the soldiers under his command: "Many of the older and better qualified non-commissioned officers had been commissioned as temporary officers.... Thus, I was to find myself with four young reserve lieutenants, a first sergeant of weak and timid character, other non-commissioned officers of limited experience, and some two hundred privates of less than a year's service."[32]

General McMahon's "Red Diamond" Division began Pershing's three-month training regimen in April 1918 at AEF Training Area Thirteen near Bar-sur-Aube, France. Officers of the advance training detachment rejoined their units as instructors at this time. For approximately one month, the division's infantry brigades learned basic trench warfare tactics, such as defensive chemical warfare techniques and trench raiding. Lieutenant Olsmith observed how trench warfare training

IV. AEF Combined Arms and Open Warfare in Action

in France differed from his previous instruction: "The training of troops for trench warfare, as prescribed in the training manuals of the American Expeditionary Forces, differed materially from that with which I had been familiar, my own training in the new type of warfare proceeded concurrently with that of my company."[33]

The Fifth Field Artillery Brigade remained at LeValdahon for its instruction. The Thirteenth, Fourteenth, and Fifteenth machine-gun battalions trained around Bar-sur-Aube, but did so separately from the infantry. Training divisional elements separately from one another was typical of all U.S. Army divisions during the Great War and greatly detracted from their ability to conduct cohesive combined arms tactics in combat. For example, American machine-gun crews never provided fire support for live infantrymen during their training exercises.

The first phase of the Fifth Division's training in France ended on 31 May, when General Pershing ordered the division to a "quiet" zone on the western front near the town of Épinal in the Toul Sector. The Vosges Mountains stretched across much of this area and inhibited large-scale military activity. This sector of the front saw only occasional patrols, raids, and minor exchanges of artillery fire. Long periods of inactivity allowed the French Army to incorporate permanent steel-reinforced concrete defenses into their trench system around Épinal. These structures were not typical of the dirt, wood, and sandbag construction of the French trenches in heavy combat zones. This represented another way in which the Fifth Division's training experience failed to match the combat realities it would later face.

The division's Ninth and Tenth infantry brigades joined with two divisions of the XXXIII Corps of the French Seventh Army for this phase of instruction. French officers taught American enlisted men trench construction, defense against poison gas, patrolling, and raiding. These raiding exercises represented the division's first opportunity to handle live grenades. In keeping with General Pershing's emphasis on marksmanship, the Fifth Division's brigade commanders insisted that the French training officers include target practice in their trench warfare regimen. *The Official History of the Fifth Division USA* recounted, "The Americans still clung to the idea that the rifle was the main dependence in warfare, and pushed training with that arm to the utmost."

The entire Fifth Division moved to the Anould Sector near Lorraine on 14 June and joined the French Twenty-First Division for more "quiet" sector training. The division suffered its first 120 casualties while serving with the French in this area. Ultimately, the Fifth's time with the French Army included very little combined arms and open warfare training. Nevertheless, Corporal John H. Smith, a machine gunner in the Sixtieth Regiment of the Ninth Infantry Brigade, showed some understanding of his combined arms role in a letter he wrote on 6 July, "A machine gunner's job is to protect the infantry and it is important work and must be done coolly and accurately."[34] Ironically, Smith had yet to actually operate a machine gun in combat at that time.

The third and final stage of the Fifth Division's training barely resembled the month of open warfare instruction General Pershing intended. Instead, the division occupied its own position on the western front in the Saint Die Sector near Lorraine on 19 July. Its first independent action took place at 4:00 a.m. on 17 August near Frapelle. After a ten-minute preliminary bombardment, the Third Battalion of the Sixth Regiment from Brigadier General Walter H. Gordon's Tenth Infantry Brigade successfully advanced into the Fave River Valley and captured the town of Frapelle. Surprisingly, this short operation included some successful combined arms. The infantry moved forward behind an effective creeping barrage provided by batteries from Brigadier General Clement A. Flagler's Fifth Field Artillery Brigade. Olsmith, a major in the Sixth Infantry Regiment by this time, described the infantry assault as "strongly supported by artillery, mortars, machine guns and American aviation." The demands of the modern industrialized battlefield, survival instinct, and experiential learning were already forcing the Fifth Division to employ advanced tactics beyond the scope of their formal training. The month spent in the Saint Die Sector cost the Fifth Division about six hundred casualties. Pershing officially ended the Fifth Division's training period on 4 September and sent the division to participate in the Saint Mihiel Offensive.[35]

That September, Harry M. Barthel of K Company of the Sixth Regiment of the Tenth Infantry Brigade filled a pocket notebook with all he had learned about weaponry and tactics through training by that time. He wrote about how French instructors criticized the AEF's

IV. AEF Combined Arms and Open Warfare in Action

dependence on the rifle and urged American trainees to rely on grenades instead. Despite this, Barthel remained faithful to his weapon, claiming that the "rifle found to be the principle weapon of the infantry" and the grenade "should not replace the rifle, use the rifle where it can be used and the hand grenade for cleaning up." Barthel came to understand the importance of coordination with the artillery. He wrote, "Get into liaison with that arm of the service; always report the formation of the front line as you advance so a protection barrage can fall in front of you." Barthel also recognized the value of suppressing machine-gun fire during infantry assaults. He even mentioned the importance of infantry and tanks advancing together, despite having no personal experience fighting alongside tanks. Finally, Barthel's notes described the open warfare application of automatic rifles in both suppressing and flanking enemy defensive positions: "The automatic rifle is efficient because it is a small target and equivalent to a large group of infantry ... and ... easily maneuvered."[36]

Barthel used the term "gangs" to refer to a "new group" on the battlefield. A "gang" consisted of one rifleman, one automatic rifleman, one grenadier, and one rifle grenadier. This "self supporting" combat formation resembled and functioned like the squads that the U.S. Army would deploy in future wars. Barthel highlighted the "three tactical principles" he had come to value by September 1918, combining artillery fire with infantry maneuver, choosing the appropriate weapon for a given objective, and "teamplay" between all arms.[37]

With its formal training completed, the Fifth Division participated in the Saint Mihiel Offensive on 12 September. It was here where the division began tactical learning through combat experience and battlefield survival. General McMahon had several combined arms components at his disposal during the offensive, including sixty-three tanks, the Twelfth Aero Squadron, the Second Balloon Company, and several units from the First Gas Regiment. After a four-hour preliminary bombardment, the Sixth and Eleventh infantry regiments of the Tenth Brigade attacked at 5:00 a.m. under a creeping barrage. First Sergeant Clyde Heldreth of D Company of the Sixtieth Infantry Regiment noted the effectiveness of this creeping barrage: "At 5:00 o'clock came the command 'over the top.' The artillery bombardment then changed to a rolling barrage. Our artillery shelling had accomplished the desired results and the enemy was in full retreat." The AEF's former chief of training, Brigadier General Paul Malone, was the Tenth Infantry Brigade's commander at the time. By 1:30 p.m., Malone's infantry advanced beyond the range of its artillery cover and outpaced its tank escort. The brigade managed to capture its objective, Bois de la Rappe, despite the breakdown of coordination between arms. The Fifth Division's rapid advance was due in part to the fact that the 332nd Regiment of the Imperial German Army's Seventy-Seventh Reserve Division was in the process of withdrawing from Saint Mihiel when the Americans attacked. Private Arbuckle noted, "The Huns were going so fast they didn't have time to use their artillery on us." The German Thirty-First and 123rd divisions relieved the depleted Seventy-Seventh and halted the Fifth Division's advance near Rembercourt later in the afternoon on 12 September. The division's Sixth and Eleventh infantry regiments secured Bonvaux and Bois Hanido before General Pershing ended the Saint Mihiel Offensive on the night of 14 September. The Fifth Division suffered nearly sixteen hundred casualties during the engagement.

The Fifth Division's participation in the Meuse-Argonne Offensive from 12 to 22 October represented the bloodiest phase in its experiential learning, costing nearly one quarter of its original strength. The Fifth's initial strategic objective was to help General Hines' III Corp break the German Kriemhilde Line by clearing enemy forces from Bois des Rappes. German artillery positions in the surrounding towns and on the heights along the eastern bank of the Meuse River protected the strong enemy defenses in these thick woods. The Fifth Division bore the additional burden of going into this battle without its organic artillery brigade. After Saint Mihiel, General Pershing detached the Fifth Field Artillery Brigade from the Fifth Division and reassigned it to the Seventy-Eighth Division.[38]

The operation started on a positive note, when the Fifth Division's Lieutenant Samuel Woodfill singlehandedly neutralized four German machine-gun positions in the Bois de la Pultiere near Cunel on 12 October. Lieutenant Woodfill received the Medal of Honor for this action. The attack took a turn for the worse over the next four days. After a two-hour preliminary bombardment, and with the support of a creeping

IV. AEF Combined Arms and Open Warfare in Action

Ninth Brigade Commander Brigadier General Joseph C. Castner (front row, fifth from left), Fifth Division Commander Major General Hanson Ely (front row, sixth from left), and Tenth Brigade Commander Brigadier General Paul B. Malone (front row, seventh from left) at Fifth Division Headquarters in Cunel.

barrage, provided by field artillery regiments from the Eightieth Division, the Ninth and Tenth infantry brigades attacked Hill 260, Hill 271, and Bois de la Pultiere at 8:30 a.m. on 14 October. This assault quickly degenerated into a defensive action when the Germans launched strong counterattacks from Bantheville, Romagne, and Cunel. Brigadier General Joseph C. Castner's Ninth Brigade attacked Bois de la Pultiere and Bois de Rappes the next morning. The American infantry, however, went in a half hour behind schedule and fell far behind its creeping barrage. The brigade managed to clear Bois de la Pultiere, but could not penetrate Bois des Rappes without artillery support. Generals Hines and Pershing quickly grew frustrated over the division's heavy losses and lack of progress and determined that the Fifth Division required a change in leadership. Major General Hanson Ely assumed command of the "Red Diamond" Division on 17 October.[39]

General Malone's Tenth Infantry Brigade finally captured Bois des

Rappes on 21 October. He created the element of surprise by launching his assault after an intense, but brief, five-minute preliminary bombardment. The infantry moved forward at 11:30 a.m. under a creeping barrage and cleared the woods of German defenders by 5:30 p.m. The attack on Bois des Rappes represented the division's last combat lesson before the final push of the Meuse-Argonne Offensive on 1 November. General Pershing pulled General Ely's division off the line on 22 October for rest and resupply. This phase of the Meuse-Argonne Offensive had cost the Fifth Division nearly forty-five hundred casualties.[40]

The Fifth Division returned to the Meuse-Argonne Sector on 26 October and joined General Liggett's First Army in regrouping and refitting before executing his plan to cross the Meuse River and capture Sedan. Corporal Carl Noble, who tended supply horses in the Sixtieth Infantry Regiment, noticed the buildup of ammunition and materiel as his unit moved to the front: "There were fields of shells, large and small.... There were trainloads of rations, stacked up in great piles."[41]

The Fifth Division incorporated combined arms into its scouting patrols prior to the assault. General Ely's *Field Orders Number 63* directed the artillery to support patrols from General Castner's Ninth Infantry Brigade into the wooded terrain of Bois de Babiemont. *Annex Number 1 to Field Orders Number 63* arranged for aircraft from the French 284th Aero Squadron to provide reconnaissance for theses patrols. These observation pilots used wireless radio to notified the artillery to cease fire when Castner's troops reached the woods.[42]

As per General Ely's orders, General Castner sent patrols into Bois de Babiemont on 28 October. *Operation Memorandum Number 119* described the combined arms aspects of this operation. The Third Artillery Brigade conducted a two-hour preliminary bombardment of the woods. Just as Ely had specified, the barrage did not cease until the patrols reached the tree line. The artillery successfully maintained wireless radio contact with the French observation aircraft during this bombardment. The Seventh Engineers assisted the Third Artillery with moving their seventy-five-millimeter guns forward as the patrols advanced to maximize the effectiveness of their fire. The patrols also received overhead machine-gun fire support from the Fourteenth Machine-Gun Battalion as they approached the woods. Once inside

IV. AEF Combined Arms and Open Warfare in Action

the forest, the scouting parties employed automatic rifles and thirty-seven-millimeter guns against enemy defenders.[43] The patrols into Bois de Babiemont on 28 October represented a microcosm of General Ely's combined arms plan for the main assault four days later.

Company F of the Sixty-First Regiment of General Castner's Ninth Infantry Brigade launched a limited attack against Aincreville on 30 October. This minor prelude to the final phase of the Meuse-Argonne Offensive also featured combined arms. The infantry waded across the Andon River and signaled for support from the Fourteenth Machine-Gun Battalion with a green flare when it reached the east bank. This prompted suppressing machine-gun fire as the infantry approached the village. Upon reaching Aincreville, a second flare signaled the machine-gun crews to shift their fire to the edges of town to prevent the Germans from retreating.[44]

The last day of October represented General Ely's final opportunity to plan for the big push. His *Field Orders Number 65* prepared the engineers for a major combined arms operation. Ely placed a company of engineers at the disposal of both infantry brigade commanders. His orders stated, "One company of engineers will be attached to each infantry brigade, and will be used to facilitate its advance by opening roads and trails." He stressed the role engineers would play in getting the division across the Meuse River: "The Division Engineers will construct foot bridges for the passage of the infantry, and, ... will construct a pontoon bridge in the vicinity of LINY-devant-DUN, suitable for the passage of 75's. A bridge for heavy artillery will be constructed at DUN-sur-MEUSE."[45]

Annex Number 2 to Field Orders Number 63 outlined General Ely's communication and liaison plans for the upcoming combined arms assault. His *Plan of Telephone Liaison* arranged for communication redundancy in anticipation of enemy artillery fire severing telephone wires during the attack. Ely required both brigades to maintain at least two telephone connections to division headquarters. He ordered all communication within four miles of the combat zone be coded in case of enemy wiretapping. Ely's *Plan of Radio Liaison* outlined details for wireless radio links between observation aircraft and regimental headquarters. Regimental headquarters communicated all pertinent information directly to their supporting artillery battalions. Unlike the artillery, attacking infantry units had no direct radio contact with reconnaissance aircraft. Ely's *Annex Number 2* recommended infantry units and observation pilots use flare pistols to communicate. Infantry units disclosed their location on the ground with a flare and pilots would acknowledge them with a response flare.[46]

Fifth Division's Combined Arms in Action, 1–11 November 1918

Wartime field orders, the official history of the division, and the memoirs of its officers and enlisted men revealed that, despite insufficient training, the Fifth "Red Diamond" Division learned through combat experience and performed combined arms and open warfare tactics during the final phase of the Meuse-Argonne Offensive. The Third Artillery Brigade began its preliminary bombardment at 3:30 a.m. on 1 November. Two hours later, two companies of the Sixty-First Regiment of General Castner's Ninth Infantry Brigade advanced toward Bois de Babiemont behind a creeping barrage and with overhead machine-gun fire support from the Fourteenth Machine-Gun Battalion. Intense German artillery fire halted the American attack just outside the tree line of Bois de Babiemont around 10:00 a.m. The German Eighty-Eighth Division had been defending this sector since 21 October. The Intelligence Section of the AEF's General Staff described this enemy division as "very fair quality and well-disciplined." Meanwhile, the Sixtieth Regiment of Castner's infantry brigade crossed the Andon River and drove the Fifth Bavarian Reserve Division from the village of Cléry-le-Grand with machine-gun support from the Fourteenth Machine-Gun Battalion. Aerial reconnaissance reports revealed that the attacks of 1 November broke the Kriemhilde Line and the Germans were forming a new defensive position along the eastern bank of the Meuse River.[47]

The critical role of the Seventh Engineers Regiment in General Ely's combined arms operation began on 3 November as the Fifth Division pushed across the Meuse River. The river was just twenty-five yards wide and five feet deep in this region, but its strong current made

IV. AEF Combined Arms and Open Warfare in Action

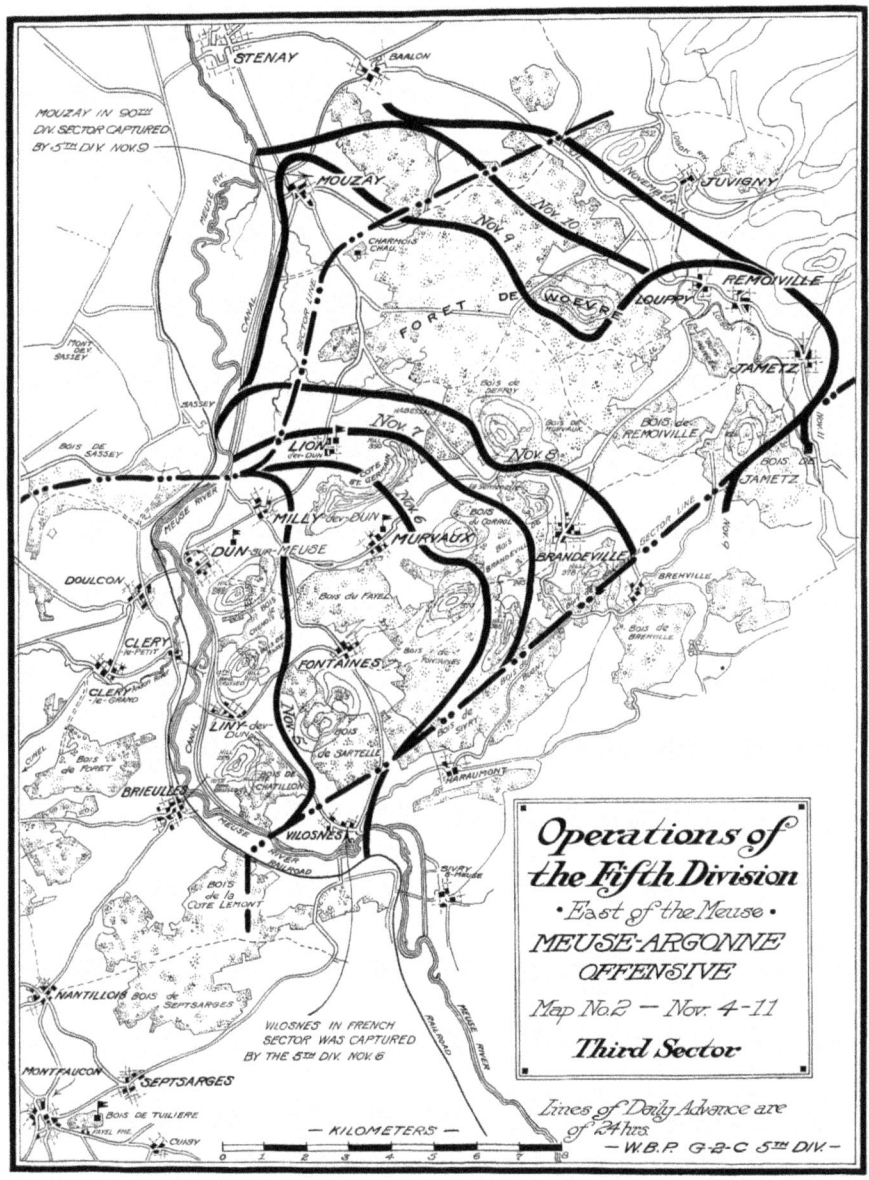

The Seventh Engineers Regiment constructed this footbridge across the Meuse River on 3 November 1918 at Brieulles. Troops of the Sixth Regiment of the Fifth Division's Tenth Infantry Brigade were the first American soldiers to cross the Meuse.

fording impossible. The canal running parallel to the river's east bank represented an additional obstacle. The canal was just five feet deep, but stone embankments on both sides extended nearly twelve feet above the waterline. The Seventh Engineers built a footbridge across the Meuse River near Brieulles during the early hours of 3 November. Darkness and rain shrouded the engineers from the German machine guns on the heights along the opposite riverbank. Around 1:00 a.m., twenty men from the Sixth Regiment of General Malone's Tenth Infantry Brigade became the first Americans to cross the Meuse River. Dense enemy machine-gun fire came with the break of dawn and forced the engineers to abandon the construction of a footbridge across the canal. The engineers relocated about half a mile down the canal to build a new footbridge beyond the range of the German machine guns. In the meantime, General Ely's *Field Orders Number 68* called for the Seventh Engineers to construct a large pontoon bridge to accommodate trucks, wagons, and artillery across the Meuse River at Brieulles for Tenth Brigade and another at Dun for the Ninth Brigade.[48]

IV. AEF Combined Arms and Open Warfare in Action

Horse-drawn artillery of the Third Artillery Brigade cross a large pontoon bridge over the Meuse River at Dun around 6 November 1918. The Seventh Engineers' construction of this bridge on 5 November was critical to keeping the Fifth Division's artillery support within range of the advancing infantry.

The engineers finally completed a footbridge across the canal just after midnight on 4 November. German machine guns repulsed two companies from the Sixth Infantry Regiment when they tried to cross the canal around 2:00 a.m. Throughout the next day, troops of the Sixth Infantry Regiment directed a variety of special weapons, including automatic rifles, rifle grenades, and Stokes mortars, against the enemy machine-gun positions, but failed to neutralize them. Meanwhile, German machine-guns of the Fifth Bavarian Reserve Division forced the Seventh Engineers to abandon their work on the pontoon bridge at Dun. The bullet-riddled bridge eventually sank into the Meuse River, taking seventy-six engineers with it. With no way to cross the river, General Castner's Ninth Infantry Brigade was stuck on the western bank of the Meuse unable to support the Tenth Brigade in its fight to cross the canal.

The Sixth Regiment of the Tenth Infantry Brigade finally crossed the canal around 5:00 p.m. on 4 November with fire support from the Tenth Artillery Brigade. For over an hour, troops of the Sixth Regiment poured across the footbridge as American artillery fire from the western side of the canal suppressed the German machine guns on the opposite bank. Some shells from the Tenth Artillery fell short and accidentally damaged the footbridge causing a temporary interruption of the crossing. *The Official History of the Fifth Division USA* described the incident: "For some unaccountable reason, just as the crossing had been successfully accomplished, shots from our own heavy ... batteries began falling short and struck all around the bridgehead, causing casualties." Despite this setback, General Malone's entire brigade crossed the canal before the next day. The Fifth Division achieved a strategic milestone by crossing the river and canal on 4 November, but the day also brought logistical problems. General Ely's division had outrun both its artillery support and supply lines. Corporal Noble described how German artillery fire also complicated American logistics: "When supplies were being taken up to the front during the daytime, the wagon train was usually shelled by the enemy. Sometimes the wagoners would run their teams through a barrage, and nearly always an animal would fall or get tangled up one way or another, and the delay exposed us to the barrage longer than if we had taken things more slowly."[49]

The next day began much like the previous. The Seventh Engineers completed two footbridges near Dun after midnight on 5 November and moved on to begin working on a canal footbridge. When the sun rose around 5:00 a.m., the German Fifty-Sixth Machine-Gun Detachment on Hill 260 opened fire on the engineers just as they finished building the footbridge across the canal. General Castner's Ninth Brigade came under intense machine-gun fire as they made their way over the canal. The footbridge eventually collapsed, and an entire company had to wade across the canal and scale the stone embankment. Captain Edward Allworth received the Medal of Honor for leading his company across the canal and capturing nearly one hundred German prisoners. With both regiments across the river and canal, Castner's Ninth Brigade neutralized the German machine guns on Hill 260 and pressed toward the towns of Dun and Milly around 8 a.m. Castner's troops employed both infiltration tactics and combined arms along the way. His lead battalions bypassed enemy strong points in Dun and left them for the support battalions. The Fifth Division's official history

IV. AEF Combined Arms and Open Warfare in Action

The Seventh Engineers Regiment construct a footbridge across the canal east of the Meuse River. The Fifth Division's Tenth Infantry Brigade crossed this footbridge under heavy enemy fire on 4 November 1918.

described the operation: "The lines merely passed over the territory, leaving mopping-up to the succeeding lines." General Castner placed his own machine guns on Hill 260 to cover the Sixty-First Regiment's approach to Milly. The division's official history recounted this combined arms affair: "Artillery from the hills farther east covered the advance while machine guns hidden ... on steep Côte St. Germain sowed the road with bullets."[50]

Meanwhile, the Seventh Engineers completed a large pontoon bridge across the river near Dun. This bridge was wide and sturdy enough to accommodate the trucks and wagons necessary to service the needy units east of the river. For example, Ambulance Company Thirty crossed this bridge on 6 November and set up aid stations near Dun and Milly. The previous five days had cost the Fifth Division over six hundred casualties, at least 80 percent of which were wounded. The pontoon bridge also enabled the division to conduct more effective combined arms, as artillery batteries crossed the river and provided the infantry with more accurate fire support. General Ely's *Field Orders*

American Tactical Advancement in World War I

The "Red Diamond" Division's Tenth Infantry Brigade crosses a pontoon bridge over the Meuse River constructed by the Seventh Engineers Regiment around Brieulles.

Number 71 directed all of his artillery to move across the Meuse River as soon as possible. By 6 November, most elements of the Fifth Division east of the Meuse had been without artillery support for two day.[51]

The sixth day of November began with the Sixtieth Infantry Regiment's capture of Côte Saint Germain. Although machine guns covered their advance, taking the heavily defended hill cost nearly thirty casualties. The work of the Seventh Engineers continued, as it constructed a wooden plank road over the quarter of a mile of swampy terrain between the Meuse River and the canal near Dun. The engineers completed this project under constant German machine-gun fire and an aerial bombardment. This road assisted in the movement of trucks, wagons, and artillery over the muddy ground that had bogged down the infantry the previous day. The Tenth Artillery crossed the Meuse River, traversed this roadway to the canal, and massed near the town of Fontaines. Despite the Tenth Artillery's progress, over-extended ammunition lines interfered with their ability to give immediate

IV. AEF Combined Arms and Open Warfare in Action

support to the infantry. The Fifth Division's artillery ammunition depots were still over eleven miles west of the Meuse. The division's official history explained, "Due to the very rapid progress of the infantry and the roughness of the roads, little use could be made of the artillery." The Eighty-Eighth Aero Squadron's inability to provide the division with reconnaissance updates due to fog and rain compounded the difficulties of 6 November. These problems ultimately led General Ely to temporarily halt the advance of his division to allow supplies and ammunition to be brought forward. Corporal Noble's memoirs reflected on the night of 6 November as a time when "the kitchen and the ration and water carts were moved up to the infantry." The division's official history claimed: "The men of the Red Diamond were weary and hungry and worn by the advance that had been so rapid as to leave supplies far in the rear, by the rough country that had confronted them every step of the way since the crossing of the river, by the rain that seemed perpetual and by the cold of early winter."[52] The Fifth Division had advanced nearly six miles since 1 November and captured over two hundred German prisoners.

The "Red Diamond" Division spent 7, 8, and 9 November consolidating its gains east of the Meuse River and patrolling the hills and wooded terrain around the French towns of Brandeville, Remoiville, Murvaux, Mouzay, and Lyon. German machine guns on the surrounding heights constantly harassed these scouting patrols. Enemy machine-gun fire also made it difficult for the Ninth Signal Battalion to maintain the telephone wires connecting the Ninth and Tenth brigades with General Ely's division headquarters at Cunel.

The Fifth Division also used these three days to regroup, refit, and shorten its over-extended supply lines. Private John Cherpak's battle account of 6 November illustrated the dire need for replacements in the vanguard of the Fifth Division. There was a particular shortage of special weapons operators. Cherpak recalled a sergeant from A Company of the Sixtieth Regiment seeking volunteers to carry the automatic rifle for an upcoming patrol, as all of the company's experienced automatic riflemen had been killed or wounded since 1 November. Sergeant Heldreth of D Company of the Sixtieth Regiment finally received rations on 8 November: "About 8 o'clock this same morning, the ration

The Third Artillery Brigade conceals a French Schneider 155-millimeter howitzer from enemy aircraft. General Ely arranged for 155s to provide heavy artillery support for Fifth Division attacks on Montmédy and Longuyon on 11 November 1918. The armistice interrupted that plan.

details reached the company with cooked rations, the first rations that had arrived since Nov. 5."[53]

General Ely issued *Secret Field Orders Number 73* at noon on 9 November calling for the division to resume its combined arms attack on 10 November. He designated Juvigny as the Ninth Brigade's objective, as the wide roads leading to the town would enable trucks, wagons, and artillery to move forward with the infantry. The town of Jametz was Tenth Brigade's objective. Jametz was an ideal location for crossing the next major topographical obstacle, the Loison River. Ely assigned a regiment of artillery and a company of engineers to each brigade. He instructed the artillery to fire on Juvigny and Jametz until the infantry entered the towns. He ordered the engineers to keep the roads passable for trucks, wagons, and artillery.

Corporal Noble noted the importance of the engineers during the Fifth Division's 10 November advance: "The Thirteenth Machine-Gun Battalion, motorized with Ford trucks" was "unable to go farther until the road was repaired."[54] General Malone's Tenth Brigade captured

IV. AEF Combined Arms and Open Warfare in Action

Jametz from the German Seventy-Fifth Division by nightfall on 10 November and took nearly sixty prisoners in the process. Moving into Jametz placed the division less than one hundred miles from the German border. The Seventh Engineers began building a bridge across the Loison River that night. The construction noises attracted German machine-gun fire from across the river, but the engineers completed the bridge. General Ely's *Secret Field Orders Number 75* set the French towns of Montmédy and Longuyon as the strategic objectives for 11 November. Ely planned to expand his combined arms assault with heavier artillery support and chemical agents. He arranged for a battalion of 155-millimeter howitzers to provide the infantry with indirect fire support as they attacked Montmédy and Longuyon. Sergeant Major Paul A. Doty of Fifth Division Headquarters observed the effects of this heavy artillery barrage: "saw a good many signs that showed the Germans had left in a hurry. Many abandoned ammunition and supplies ... judging by the number of shell holes ... they must have been given a hearty send-off by the American artillery." Ely invited the First Gas Regiment to assist the Fifth Division on 11 November as well. At approximately 8:00 a.m., however, General Ely canceled the attack and notified his brigade commanders of the armistice. The cost of the final phase of the Meuse-Argonne Offensive increased during its final days. The Fifth Division suffered nearly eleven hundred casualties between 6 and 11 November.[55] Fortunately, the Great War was over for the AEF and its "Red Diamond" Division.

Conclusion

Consistent with the majority of AEF divisions, formal training did little to prepare the soldiers of the Fifth Division to conduct combined arms and open warfare in combat. All phases of the "Red Diamond" Division's instruction were rushed, and there were few weapons available for training purposes. Furthermore, the division's various combat arms trained in different locations around France. Based upon the Fifth Division's performance during the final phase of the Meuse-Argonne Offensive, it was six months of combat experience and survival on the western front that allowed it, and the rest of First Army, to conduct

American Tactical Advancement in World War I

Top: Troops of the "Red Diamond" Division bury the fallen. The Fifth Division sustained more than eighteen hundred casualties during the final phase of the Meuse-Argonne Offensive from 1 to 11 November 1918. *Bottom:* Cease-fire 11 November 1918.

IV. AEF Combined Arms and Open Warfare in Action

combined arms and open warfare to the point of achieving a successful breakthrough of the German Kriemhilde Line. Deficiencies in the areas of supply, logistics, and communications prevented the AEF from exploiting this breach to the point of annihilating the Imperial German Army. These underdeveloped support systems forced First Army and its Fifth Division to pause and refit before crossing into Germany. This respite may have allowed the Imperial German Army to establish a strong line of defense along the Rhine River and drag the conflict into 1919. According to Colonel Lanza: "The Armistice put an end to what would have been a long wait, possibly extending through the winter into the following spring. In other words, the Allies managed to give the enemy the final blow just before it would have been necessary to stop the offensive to reorganize and reequip."[56]

Conclusion:
U.S. Army Combined Arms and Open Warfare After the Great War

After the armistice, the U.S. Army continued to develop the tactical lessons learned by the American Expeditionary Forces on the western front. Combined arms and General Pershing's open warfare concept were persistent themes in army doctrine through the 1920s. Supporting the infantry with close artillery and machine-gun fire were standard practice. The army continued working to define and expand combined arms roles for tanks and aircraft. American military thinkers carried on the conversation over trench-warfare versus open warfare tactics. By the 1930s, however, diplomatic isolationism and the budgetary limitations of the Great Depression interrupted the army's tactical growth. Under these restraints, U.S. Army weapons and equipment became antiquated and costly large-scale training maneuvers were nonexistent. The lack of funding for either improved weapons or training gradually led to a state of under preparedness in everything but thought. Combat experience was, once again, the army's primary method of learning to survive and succeed in battle during the Second World War. The U.S. Army of 1941 was not unlike the AEF in 1917 in this regard. Its greatest advantage resided in the leadership of veteran commanders, who possessed the wisdom of the Great War's tactical learning curve.

Interwar Combined Arms Doctrine and Training

Immediately following the Great War, the U.S. Army studied the AEF's experiences to identify the tactical lessons of the conflict and gauge the future direction of warfare. General Pershing, now General of the Armies of the United States, was enthusiastic about the role of new technology in the next war. As a result, combined arms became a standard feature of American tactical doctrine in the 1920s. West Point made combined arms a permanent part of its curriculum in 1919 under new superintendent Brigadier General Douglas MacArthur. MacArthur had commanded a brigade in the Forty-Second Division during the Meuse-Argonne Offensive and was the most decorated officer in the AEF. The U.S. Army Command and General Staff College adopted combined arms into its curriculum as well. In 1919, Chief of Staff Peyton March persuaded the government to pass a budget permitting the army to conduct large-scale, combined arms, training maneuvers in the United States.[1]

The U.S. Army's 1921 *Principles of War* identified cooperation between arms as a key tenet of modern industrialized warfare. The War Department published *War Department Training Regulations Number 10–15* that same year. It contained guidelines for coordination between all arms. The most influential document reflecting the U.S. Army's lessons from the Great War was *Field Service Regulations 1923*. The War Department General Staff and the faculty of the U.S. Army Command and General Staff College based this manual on the AEF's combat experiences. The publication detailed sophisticated combined arms operations involving infantry, artillery, tanks, and airplanes. Furthermore, the manual stated that every American infantryman should be familiar with the basic applications of artillery, chemical weapons, machine guns, and tanks. Rather than cast the infantry as simply riflemen, as pre–Great War American doctrine had done, *Field Service*

Conclusion

Regulations 1923 described the infantry as a combined arms force equipped with mortars, grenades, automatic rifles, and thirty-seven-millimeter cannon.[2]

The U.S. Army Infantry School at Fort Benning, Georgia, conducted its first combined arms exercise in 1923 with elements of infantry, artillery, chemical troops, a tank company, some engineers, and even mounted cavalry. Colonel George Marshall became assistant commandant of the school in 1927 with Major Joseph W. Stilwell in charge of tactical instruction. Stilwell had served as an intelligence officer in the AEF's IV Corps. Marshall and Stilwell encouraged their officer trainees to consider all weapons at their disposal when solving battlefield problems. Marshall coordinated a large-scale combined arms exercise in 1927, including the Second Infantry Division, the First Cavalry Brigade, and two hundred aircraft.[3]

The U.S. Army carried out all of its training in the 1920s with weapons and vehicles left over from the Great War. Congress believed that the vast surplus of equipment from 1918 did not justify an army budget increase. This posted a problem when weapons technology from the Great War became antiquated by the late 1920s. Needless to say, the U.S. Army's budget and equipment dilemma only worsened during the Great Depression. For example, Colonel Marshall just barely commandeered enough serviceable weapons and vehicles for another large-scale combined arms maneuver at Fort Benning in 1932. The U.S. Army General Staff attempted to update its arsenal in the mid–1930s. General MacArthur, now the army's chief of staff, published his *Annual Report of the Chief of Staff* in 1934. He estimated that the U.S. Army, with under 140,000 troops and just twelve tanks, ranked only seventeenth among the armies of the world. MacArthur, therefore, requested the addition of thirty-seven hundred modern artillery pieces, sixty-seven new tanks, enough vehicles to motorize the entire service, and the replacement of the bolt action 1903 Springfield with a semiautomatic rifle. Great Depression budget cuts forced Congress to suspend MacArthur's weapons development plan.[4] As a result of these Depression-era spending limitations, the U.S. Army still equipped its soldiers with vintage seventy-five-millimeter cannon, Stokes mortars, 1903 Springfield rifles, and Renault tanks when the Second World War erupted in 1939.

U.S. Army Combined Arms and Open Warfare After the Great War

The U.S. Army experienced a manpower shortage in the 1930s as well. This too was a result of Depression-era financial restrictions, but also connected to the government's return to a diplomatic policy of neutrality and isolation between the world wars. The Kellogg-Briand Pact of 1928 reflected America's posture on international affairs. This agreement involved the United States and sixty-one other nations pledging to settle their future diplomatic differences through peaceful negotiation. Even the War Department preoccupied itself with defensive strategic planning in the 1930s, such as defending against hypothetical Mexican invasions of Texas and British amphibious landings on the eastern seaboard.[5]

The War Department promptly scaled down the U.S. Army to approximately five hundred thousand men after the Great War. The National Defense Act of 1920 reduced this to three hundred thousand. In actuality, the U.S. Army did not number over 150,000 officers and enlisted men for most of the 1930s. Twenty-five percent of these troops served outside the United States, where they would be exposed to none of the tactical developments and training happening stateside. Needless to say, large-scale training exercises were a rarity during the interwar period. Most army training maneuvers in the 1920s and 1930s involved nothing larger than a battalion.[6]

Without the ability to carry out large-scale training exercises with the latest technology, the U.S. Army's combined arms development stagnated during the ten years leading up to the Second World War. The War Department General Staff's *Field Service Regulations 1939* showed only minor growth since *Field Service Regulations 1923*. It provided for special weapons units, such as automatic-rifle, mortar, and machine-gun teams, to be integrated into infantry platoons and companies.[7] These new formations did not materialize, however, until the army actually entered the Second World War. The U.S. Army of 1941 looked similar to its 1917 counterpart, grossly unprepared for the coming conflict. The deserts of North Africa and islands of the Southwest Pacific would serve as its combined arms classroom in 1942.

Conclusion

Interwar Artillery Doctrine and Training

The Great War convinced American military thinkers that infantry could not advance across the modern industrialized battlefield without effective, close, artillery support. Conrad Lanza wrote an article in *The Field Artillery Journal* in 1935 stating that "it was evident that the enemy had been demoralized by our artillery preparation, and particularly by the heavy rolling barrage."[8] The U.S. Army planned to expand upon the lessons of the war with a robust combined, artillery/infantry, training regimen in the 1920s. American artillery also sought to modernize its arsenal during the interwar period with full motorization, wider use of wireless radio communication, and new weapons systems. Unfortunately, Depression-era budget restrictions limited all attempts to update U.S. Army artillery tactics and technology between the world wars.

While serving as senior instructor to the Illinois National Guard in 1934, Colonel Marshall oversaw the publication of *Infantry in Battle*. The book recognized artillery as the infantry's most important support arm. According to the publication, the artillery should bombard enemy positions prior to an attack and provide suppressing fire until the infantry closes to within one hundred yards of the objective. Marshall asserted that forward infantry units needed to keep the artillery abreast of critical battlefield data, such as target type, target location, and the position of friendly troops. The publication noted that the AEF lacked this level of coordination during the Great War.[9] The type of close artillery support described in *Infantry in Battle* required real-time, wireless, radio, liaison between the artillery and infantry.

Colonel Marshall believed that effective combined arms depended upon artillery and infantry units establishing familiarity and mutual trust through shared training and combat experience. He criticized General Pershing's policy of separating artillery units from their parent infantry divisions during the Great War. Marshall's *Infantry in Battle* claimed that prolonged joint training between artillery and infantry units would result in near-flawless artillery support for infantry attacks, as both arms would become fully aware of each other's capabilities and limitations. As Assistant Commandant of the U.S. Army Infantry

School, Marshall established trainee and faculty exchange programs with the U.S. Army Field Artillery School at Fort Sill to encourage familiarity between arms.[10]

Meanwhile, Assistant Commandant Lieutenant Colonel Lesley J. McNair, Director of Gunnery Major Carlos B. Brewer, and gunnery instructors, Lieutenants Orlando Ward and Jacob L. Devers, worked to improve the artillery's ability to support the infantry at the U.S. Army Field Artillery School. McNair had been a brigadier general with the First Division during the Great War and Ward commanded the Second Battalion of the AEF's Tenth Field Artillery Brigade. These four artillery officers spearheaded the modernization of American artillery observation procedures in the 1930s. Recalling how communication breakdowns prevented the AEF from maintaining effective artillery support for its infantry, they appreciated the need to equip frontline artillery observers with wireless radios. Unfortunately, the artillery's restricted budget forced Colonel McNair and his subordinates to conduct their training maneuvers with only a few wireless radios. The U.S. Army still depended on wired telephones and runners in the late 1930s and did not enjoy instantaneous, wireless, communication between all artillery and infantry units until after it entered the Second World War.[11]

The U.S. Army's *Field Service Regulations 1939* remained consistent with the tactical lessons of the Great War. It continued to hold the artillery as the infantry's primary support arm. Unfortunately, *Field Service Regulations 1939* showed very little innovation in artillery tactics since the 1923 regulations. For example, the publication contained no guidelines for providing tanks with artillery support.[12]

The AEF brought six thousand artillery pieces back from Europe in 1919, raising the army's total to just over eight thousand guns. There were nineteen artillery factories in the United States during the Great War. The government shut down fifteen of these facilities after the armistice and placed five million square feet of artillery manufacturing equipment in storage. The Rochester cannon and Erie howitzer plants in New York went on "stand-by" for future operation. The Watervliet arsenal in New York and Watertown arsenal in Massachusetts remained active during the interwar years, but produced only large-caliber,

Conclusion

coastal, defense guns.[13] The manufacturing of coastal artillery was consistent with the War Department's strategic preoccupation with a possible invasion of America's eastern shore.

The U.S. Army organized several committees to conduct artillery studies immediately following the Great War. Brigadier General Andrew Hero, Jr., who commanded the 154th Field Artillery Brigade during the war, headed the Hero Board, Brigadier General William I. Westervelt chaired the Westervelt Board, and Joseph Dickman chaired the Superior Board. These committees drew conclusions concerning the technological needs of American artillery based upon the AEF's recent experience. They credited the seventy-five-millimeter cannon with ideal range, but expressed concern over its limited trajectory and small-caliber projectile. The boards praised the 155-millimeter howitzer's trajectory and caliber, but criticized its bulk and poor mobility. The U.S. Army addressed this problem of balancing range, trajectory, and caliber by adopting the multipurpose 105-milimeter howitzer in 1926. Budget restrictions, however, delayed the mass production of the 105 until 1938. As the U.S. Army started outfitting its artillery batteries with 105s, there were too many seventy-fives left over from the Great War to justify scrapping them all. As a result, the army continued using vintage seventy-five-millimeter cannon until 1943.[14]

The postwar artillery boards also identified the AEF's lack of tractors and trucks for keeping the artillery in range to support the advancing infantry as a major shortcoming. The Westervelt Board asserted that effective combined arms required all artillery pieces to be towed by motor vehicles. Once again, budget limitations hampered the artillery's technological development. The U.S. Army only managed to motorize its heavy artillery during the 1920s and 1930s. Horses continued to pull American field artillery until 1941.[15]

Interwar Chemical Warfare Doctrine and Training

After the Great War, many American politicians and military leaders hoped the international community would ban chemical agents from future wars. Ironically, the United States refrained from ratifying

Troops of the Third Artillery Brigade and Seventh Engineers Regiment struggle to move a French Schneider M1897 75-millimeter cannon without the help of horses or motor transport. The AEF's shortage of pack animals and trucks represented the artillery's greatest challenge when trying to keep up with the rapidly advancing infantry during the Meuse-Argonne Offensive.

the Geneva Gas Protocol in 1925, just as it had refused to sign the Hague Anti-Gas Declaration in 1899. Winford Lewis, the inventor of Lewisite, spent the 1920s writing articles and giving speeches in support of retaining and expanding the American chemical weapons program. Others suggested conducting scientific research in chemical agents, but suspending applied military training. Despite the mixed opinions, the National Defense Act of 1920 established an independent U.S. Army Chemical Warfare Service and a Chemical Warfare School at the Lakehurst Proving Ground in New Jersey. This arm was a continuation of the AEF's Gas Corps, with a reduced staff of fifteen hundred men. Former commander of the Gas Corps, Major General Amos Fries was chief of the new Chemical Warfare Service. *Field Service Regulations 1923* broadened U.S. Army chemical weapons tactics to include offensive applications.[16] This was an improvement upon the AEF's chemical agent manuals, whose focus were strictly defensive.

Conclusion

The U.S. Army placed its four chemical-producing facilities at the Edgewood Arsenal in Middle River, Maryland, on "stand-by" following the armistice. The arsenal produced over sixteen hundred tons of chlorine, chloropicrin, phosgene, and mustard gas during the war. After the war, the army stored fourteen hundred tons of chemicals at Edgewood. The Chemical Warfare Service at Lakehurst spent the 1930s developing new chemical weapon systems, such as bombs that could be dropped from aircraft. Meanwhile, the Italian *Regia Aeronautica* sprayed mustard gas and Lewisite on enemy troops from airplanes during Italy's invasion of Ethiopia in 1936. The War Department watched closely to see how the European belligerents would employ chemical agents in the early years of the Second World War. With no sign of chemical warfare on the battlefields of Europe, Africa, and Asia by 1941, the War Department considered dissolving the independent Chemical Warfare Service. As a result, the Chemical Warfare Service was absent from the army's training maneuvers in both 1940 and 1941. The army sold most of its surplus chemical agents commercially and disposed of twenty tons of phosgene and mustard gas off the coast of Maryland. The U.S. Army, therefore, had very few chemical agents on hand when it entered the Second World War in December 1941.[17]

Interwar Machine-Gun, Mortar, Automatic Rifle and Small Arms Doctrine and Training

Though the United States government halted the mass production of machine guns in 1919, the machine gun as an offensive combined arm became a regular feature of U.S. Army tactical doctrine during the interwar years. Colonel Marshall's *Infantry in Battle* criticized the AEF for being too slow to employ the machine gun as an offensive weapon. Marshall considered the machine gun essential for suppressing enemy positions while infantry units maneuvered to outflank. The book recognized that moving machine guns forward with the infantry presented a challenge in this regard. Machine-gun mobility remained a problem for the U.S. Army between the world wars, as the army lacked a standard light machine gun. The German and British armies both adopted mobile light machine guns by the 1930s with the German MG 34 and the British Bren Gun. The bulky Browning M2, weighing 128 pounds with its tripod, was the U.S. Army's only machine gun during the interwar period. Budget cuts prevented the army from adopting anything lighter than the M2. As a result, the U.S. Army was still without a standard light machine gun at the outset of the Second World War.[18]

Mortars and automatic rifles featured prominently in American combined arms doctrine in the 1920s and 1930s. *Field Service Regulations 1923* cast mortars as a fire-support weapon for advancing infantry. Marshall's *Infantry in Battle* described mortars reducing defensive positions prior to assaults and suppressing the enemy while the infantry attacked. The Stokes was still the army's standard mortar in the 1920s. The U.S. Army attempted to modernize its mortar arsenal in the 1930s with the M1 and M2 mortars. The M1 weighed 136 pounds and lobbed an eighty-one-millimeter projectile up to thirty-three hundred yards. The M2 fired a sixty-millimeter shell up to 1,985 yards and weighed forty-two pounds. Needless to say, Depression-era budget limits prevented the army from outfitting the entire service with these new mortars until 1941.[19]

Despite the AEF's widespread use of automatic rifles, the weapon did not appear as prominently in U.S. Army doctrine as it did with some European armies after the war. For example, the automatic rifle played a central role in French fire-and-maneuver tactics between the world wars. In the 1930s, some German tacticians predicted that every soldier would carry an automatic rifle in the next conflict. Despite the weapon's scant appearance in published American doctrine, the U.S. Army retained the automatic rifle's fire support role during the interwar years.[20] The army kept the BAR as its standard automatic rifle through the Second World War. Unlike many other American weapons of the interwar period, the BAR's longevity was due to its superior quality, not fiscal restrictions.

Years of static trench warfare on the western front caused the French and British armies to lose confidence in the infantry's ability to support itself across the battlefield with rifles alone. Both armies omitted marksmanship from their training programs by 1917. According to General Pershing, however, the AEF's experience during the Meuse-Argonne Offensive proved otherwise. He credited the American soldier

Conclusion

and his rifle with restoring mobility to the western front in 1918. After the war, Pershing believed that the rifle was still the primary means by which the infantry supported its own advance. As a result, marksmanship continued to be a prominent feature of U.S. Army tactical doctrine between the world wars. Although *Field Service Regulations 1923* included many weapons in its description of combined arms, it set the rifle as the infantry's primary means of neutralizing enemy troops and capturing ground.[21]

Interwar Tank Doctrine and Training

American tank development was at a tactical impasse when the Great War ended. General Pershing and his former AEF subordinates believed that the next war would be a mobile conflict with tanks playing a prominent role, but they disagreed over how to employ tanks. Some were confident that the tanks of the future would be capable of breaking through enemy defenses, capturing objectives, and consolidating gains as an independent arm. Others believed tanks should remain in their infantry-support role from the Great War, destroying obstacles and providing mobile cover for foot soldiers. The latter group based its conclusion on the technological limitations that afflicted tanks on the western front.[22] After all, mechanical failure was responsible for thirty-six of the thirty-eight American tanks lost at Saint Mihiel.

Major George Patton was among those who believed in the tank's potential as an independent arm after the war. He advocated strongly for the formation of an autonomous tank corps in 1919 and even presented the army with a written plan, including tactical guidelines and a training regimen, for this hypothetical tank corps. Meanwhile, the U.S. Army merged its 250 tanks from Europe with its 179 domestic tanks and sent Patton to take command of this force at Camp Meade, Maryland. For Patton, this indicated the army's intent to create his proposed, independent, tank arm.[23]

Patton drafted tank doctrine with Major Dwight Eisenhower during his time at Camp Meade. Patton and Eisenhower asserted that, while tanks were not a replacement for infantry, they could perform the same functions in combat. They envisioned massive tank attacks

U.S. Army Combined Arms and Open Warfare After the Great War

involving several waves of tanks to break through defenses, infiltrate breaches, and flank enemy positions. Patton and Eisenhower acknowledged that these sophisticated tank tactics would require combined arms support, such as aerial reconnaissance for planning attacks, artillery to neutralize enemy anti-tank threats, and infantry to consolidate captured ground.[24] Needless to say, their relegation of infantry to a support role for tanks was highly unpopular with the more traditional American military thinkers of that time.

The *Infantry Journal* published Patton and Eisenhower's ideas in 1920 in an article titled "A Tank Discussion." Their essay posited that future tanks would not be the sluggish, mechanically unreliable machines of the Great War, but precision vehicles with balanced firepower, armor, speed, and maneuverability. They encouraged infantry commanders to look at tanks as more than just a means for escorting soldiers across the battlefield and open their minds to the independent potential of tanks.[25]

Majors Patton and Eisenhower knew their vision for the future required substantial improvements in existing tank technology. Patton went to the U.S. Army Proving Ground at Aberdeen, Maryland, to research potential armament and armor upgrades for American tanks. He concluded that the best weapon configuration for tanks was one cannon and two machine guns, one mounted on the turret and the other in the hull. The hull needed multiple viewports to provide crewmen with adequate visibility and angled surfaces to deflect incoming fire. Patton determined that tanks required a minimum speed of twenty miles per hour.[26] All of these findings remained strictly academic, as a U.S. Army tank budget did not exist in 1920.

Despite Patton and Eisenhower's efforts in 1919 and 1920, the U.S. Army's opinion on the role of tanks in future wars remained mixed. Even the former Chief of the AEF Tank Corps Brigadier General Samuel Rockenbach was undecided and oddly silent during the postwar tank conversation. Ultimately, General Pershing and the War Department did not include the establishment of an independent tank arm in the National Defense Act of 1920. The National Defense Act of 1920 dispersed the army's tanks across the service, attaching fifteen tanks to each infantry division and placing them at the behest of infantry

Conclusion

commanders. Not surprisingly, the U.S. Army's published tank tactics in the 1920s resembled those of the Great War. For example, *Field Service Regulations 1923* portrayed tanks leveling obstacles and providing mobile cover for advancing infantry. Patton and Eisenhower's discussion about the tank being more than an infantry-support arm became heresy. U.S. Army Chief of Infantry Major General Charles S. Farnsworth threatened them both with court martial if they continued promoting this idea.[27]

The tank entered the 1920s as one of the U.S. Army's many infantry-support weapons, competing with other arms for limited funds. The United States spent only $1.5 million on tank technology between 1920 and 1930. With the addition of only a few new tanks, most American tanks were Mark VIIIs, Ford M1918s, and Renaults left over from the Great War. The Mark VIII was an Anglo-American heavy tank designed in 1917. Operated by a ten-man crew, the tank traveled five miles per hour. The Mark VIII's armament consisted of five Browning machine guns and two fifty-seven-millimeter cannon. A two-man crew operated the Ford M1918. The tank traveled eight miles per hour and was armed with one Browning machine gun.[28] Neither the Mark VIII nor the Ford M1918 saw combat during the Great War.

Budget was not the only limitation on U.S. Army tank development in the 1920s. The government and army set strict guidelines for American tank manufacturers. Due to the maximum weight capacity of most roads, bridges, and railroads within the United States, American-made tanks could not exceed fifteen tons. Furthermore, the army decreed that all new tanks had to have a minimum speed of at least twelve miles per hour. These regulations restricted American tank developers to making only light and medium tanks. At the same time, the army expected tank armor to repel a .50 caliber round. It was nearly impossible to produce a tank with armor thick enough to meet this standard without exceeding fifteen tons or falling short of the twelve miles per hour minimum speed. Despite the obstacles, American tank designers presented the army with twelve different prototypes in the 1920s.[29] The U.S. Army, however, did not adopt any of these designs.

An important turning point in U.S. Army tank development came in 1927, when Secretary of War Dwight F. Davis observed combined arms maneuvers in Great Britain. The British Army's independent tank force impressed Davis and inspired him to order the U.S. Army to establish an experimental mechanized unit. U.S. Army Chief of Staff General Charles Summerall formed this combined arms force at Camp Meade in 1928 with two tank battalions, one infantry battalion, one artillery battalion, a signal company, a company of engineers, a squadron of observation aircraft, and a mix of automobiles and horses. General Summerall placed Colonel Daniel Van Voorhis, Colonel Guy V. Henry, Jr., and Lieutenant Colonel Adna R. Chaffee, Jr., in command of this Experimental Mechanized Force. Henry had commanded an infantry brigade in the AEF and Chaffee served as a colonel in the AEF's III Corps. Summerall was unable to allocate funds for the unit, so it had to make due with equipment, weapons, and vehicles left over from the Great War. As a result, mechanical problems and weapon malfunctions became so frequent that General Summerall cancelled the project after only three months.[30] Nevertheless, the U.S. Army took an important step toward realizing the tank's prominent role in combined arms.

The modest strides in U.S. Army tank development in the 1920s left most American military thinkers still questioning the tank's place on the battlefield in the 1930s. Despite being momentarily inspired by the sophisticated combined arms role of tanks in the British Army in the late 1920s, American tank tactics still resembled those of the Great War for much of the 1930s. Furthermore, Depression-era budget constraints crippled an already sluggish American tank industry.

In 1930, U.S. Army Chief of Staff Douglas MacArthur established another experimental combined arms force at Fort Eustis, Virginia with twenty-three light tanks, a motorized artillery battery, a motorized machine-gun company, a company of engineers, signal and chemical agent units, eleven armored cars, thirty-three trucks, twenty automobiles, and fifteen motorcycles. General MacArthur placed Colonel Van Voorhis and Major Sereno E. Brett in command of this force. Brett had commanded the 326th Tank Battalion at Saint Mihiel in 1918. Poor funding and obsolete weaponry plagued this experimental force as well and forced MacArthur to terminate the Fort Eustis program in 1931.[31]

Without consensus on how to employ tanks in combat, the U.S. Army had no standard guidelines for tank specifications in the 1930s.

Conclusion

Chief of Infantry Major General Edward Croft concluded that lighter tanks were versatile enough to serve whatever the army's tactical needs may turn out to be in the next war. The U.S. Army, therefore, neglected heavier tank development in the 1930s. This, and Great Depression budget cuts, were responsible for the War Department's infamous decision to reject John Walter Christie's M1928 tank design in 1931.[32] Christie sold his idea to the Soviet Union, where it evolved into the T-34 Tank. The Red Army's T-34s would play a key role in driving the German Army out of Russia during the Second World War.

Patton and Eisenhower's vision from 1919 and 1920 finally found its way into mainstream U.S. Army tactical doctrine in 1938, when U.S. Army Chief of Infantry Major General George A. Lynch clarified the tank's independent function within American combined arms. General Lynch decreed that medium tanks would spearhead attacks, utilizing their firepower to destroy and break through enemy defenses. Light tanks would then use their speed and maneuverability to escort and protect the infantry as it exploits these breaches. The U.S. Army finally updated its tank technology as well. The army adopted the M2 Light Tank and the T5 Medium Tank in 1936. The M2 had a four-man crew, a top speed of thirty-six miles per hour, and was armed with one thirty-seven-millimeter gun and five .30 caliber machine guns. The T5 had a five-man crew, a top speed of twenty miles per hour, and was armed with one seventy-five-millimeter gun and eight .30 caliber machine guns.[33]

The U.S. Army published General Lynch's tank tactics in *Field Service Regulations 1939*. The manual described tanks attacking in waves. The initial wave was made up of independent medium tanks creating gaps in enemy defenses. The second wave consisted of light tanks advancing with the infantry and infiltrating enemy lines.[34] Meanwhile, the German Army unleashed seven full tank divisions, with a total of twenty-four hundred tanks, across the Polish border that same year. These *panzer* divisions breached enemy defenses, advanced deep behind enemy lines, and defended captured ground.

After several appeals before Congress in 1940, U.S. Army Chief of Staff George Marshall obtained permission to hold a large combined arms training maneuver, including tanks, motorized infantry, and mechanized cavalry, in Louisiana's Sabine River Valley. This would be the U.S. Army's first large-scale training exercise since 1912, with approximately ninety thousand troops participating. Due to weapon shortages, large pipes represented enemy artillery positions and trucks impersonated enemy tanks. Reporters from the *New York Times* observed theses maneuvers and published scathing critiques of the U.S. Army's lack of preparedness for the type of modern mechanized warfare raging overseas at the time.

The lackluster training exercise in Louisiana and the German *blitzkrieg* victory over France in the spring of 1940 convinced General Marshall to establish a new independent tank force. This U.S. Army Armored Force handled all of the army's tank development, training, and organization. The force coordinated with other arms, but remained entirely independent. The U.S. Army published its new outlook on tanks in *Field Service Regulations 1941*. The manual detailed independent tank units working alongside the infantry as mobile firepower. Six armored divisions, with over 850 tanks, and thirty infantry divisions gathered for maneuvers near Camden, South Carolina in November 1941. Secretary of War Henry Stimson met with the War Department to assess the results of this training operation just four days before Imperial Japan attacked Pearl Harbor.[35] There would be no more time to integrate new tank tactics and technology into U.S. Army combined arms before the next conflict. When the first American tank units landed in North Africa in November 1942, they were no match for the superior machines and experienced crews of the German *Afrika Korps*. It would take many months for American tankers to progress along the costly learning curve of armored warfare during the Second World War.

Interwar Mounted Cavalry Doctrine and Training

Ironically, a significant portion of U.S. Army tank development occurred within the mounted cavalry during the interwar period. After the Great War, the army questioned the need for mounted cavalry in future conflicts. American cavalry saw no combat on the western front

Conclusion

in 1918. In fact, many American cavalry officers were concerned for their careers after the war. Some, however, believed horses were still the army's best form of transportation across rugged terrain and cited the prominent role they played during General Pershing's Punitive Expedition into Mexico in 1916.[36]

The U.S. Army was divided over whether or not automobiles had rendered mounted cavalry obsolete in the 1920s. The United States was the largest automobile producer in the world when the Great War ended. Nevertheless, the army cancelled its contracts with Ford and Dodge after the armistice and sold over twenty thousand of its automobiles and motorcycles to foreign countries in 1919. At the same time, the U.S. Army sent a convoy of over seventy vehicles from Washington, D.C., to San Francisco, California to showcase the potential of military motor transport. The National Defense Act of 1920 assigned the army's armored cars to the cavalry. In October of that year, George Patton transferred to a cavalry post at Fort Myer, Virginia. With his hopes for an independent tank service stymied by General Farnsworth, Major Patton considered the cavalry to be the closest thing the army had to the kind of mobile combined arms force he and Dwight Eisenhower had endorsed.[37]

While Patton championed mechanized combined arms in the cavalry at Fort Myer in 1921, Major Bradford G. Chynoweth published an article in the *Cavalry Journal* predicting that the tanks of the future would replace mounted cavalry altogether. He believed that adopting tanks was the cavalry's only chance for retaining a place in modern industrialized warfare. Chynoweth described the armored cavalry trooper of the future as "a giant iron horseman."[38] Like Patton and Eisenhower, General Farnsworth reprimanded Major Chynoweth for suggesting that tanks would play such an independent role in future wars. Chief of Cavalry Major General Willard A. Holbrook and Secretary of War Davis, however, were enthusiastic about the mechanization of the cavalry. As a result, *Field Service Regulations 1923* omitted offensive mounted cavalry tactics and recommended horses be used only for reconnaissance. This suggested that motor vehicles and/or tanks would be the cavalry's mode of attack in the future. The cavalry conducted a massive training maneuver with armored cars along the

Mexican border in 1929. Unfortunately, most of its outdated vehicles either overheated or suffered tire damage in the rugged desert terrain.[39]

Eliminating mounted cavalry was a question of "when" rather than "if" as the U.S. Army entered the 1930s. A turning point came in 1932, when General MacArthur assigned the light tanks from the disbanded Fort Eustis experiment to the cavalry. The War Department then authorized the cavalry to establish a mechanized combined arms force of artillery, armored cars, and tanks at Fort Knox, Kentucky under the command of Colonel Chaffee. Since the National Defense Act of 1920 did not permit tanks to serve outside the infantry, the U.S. Army classified the cavalry's tanks as "combat cars."[40]

Like all other arms of the U.S. Army, the Great Depression brought budget cuts to the cavalry's combined arms development in the mid-1930s. In 1934, Chief of Cavalry Major General Leon B. Kromer petitioned the War Department for funds to outfit every cavalry division with one mechanized brigade of "combat cars." The government rejected Kromer's request. The War Department turned down a similar request by Colonel Chaffee in 1936. While economic constraints were primarily responsible for these rejections, there were also concerns within the infantry about the cavalry reaching beyond its traditional reconnaissance and pursuit responsibilities.[41]

Colonel Chaffee's force at Fort Knox, the Seventh Cavalry Brigade (Mechanized), enjoyed some technological upgrades in the late 1930s. The U.S. Army added a squadron of observation aircraft and outfitted the brigade with M1 Light Tanks("combat cars") in 1938. The M1 had a four-man crew, traveled forty-five miles per hour, and was armed with one .50 caliber machine gun and two .30 caliber machine guns. Its light armor and firepower made it a far cry from the multipurpose tanks Patton, Eisenhower, and Chynoweth had envisioned in the early 1920s. Cavalry tactics in *Field Service Regulations 1939* were not dissimilar from those of the late 1800s, restricted mainly to reconnaissance. The U.S. Army amalgamated the cavalry's "combat cars" into a new independent armored force in 1940.[42]

Conclusion

Interwar Air Support Doctrine and Training

The Great War convinced the U.S. Army that aircraft would play a major role in future conflicts. As with tanks, there were differences of opinion as to how aircraft should fit in with combined arms. After the war, generals Pershing and March believed the primary purpose of aircraft was to gather reconnaissance for the infantry and artillery. Some saw aircraft providing the infantry with close air support, strafing and bombing enemy ground targets. Still others argued that aircraft could win wars singlehandedly through long-range strategic bombardments of enemy industry and infrastructure. The War Department's tactical publications of the 1920s described aircraft in all of these capacities.[43]

The U.S. Army shipped over two thousand aircraft back to the United States after the armistice and closed its stateside aircraft factories. Only the U.S. Aeronautical Engine Plant in Long Island, New York, remained on "stand-by." The government believed that halting its aircraft production would stimulate the growth of the country's commercial aircraft industry. The army also closed twenty of its twenty-six airfields in the United States.[44]

Deputy Director of the U.S. Army Air Service William Mitchell was the strongest advocate for the formation of an independent American air force after the Great War. He believed the war had proven that achieving air superiority, gathering aerial reconnaissance, and strafing and bombing enemy targets on the ground were necessary features of a successful offensive and confirmed airpower's role as an autonomous combined arms component. He also subscribed to the notion that long-range aerial bombardments could win wars singlehandedly. Not surprisingly, Mitchell faced strong resistance from infantry commanders eager to keep airpower under their control. The National Defense Act of 1920 settled the dispute by establishing the U.S. Army Air Service as an arm within the army. General Pershing passed over the controversial General Mitchell and appointed Major General Charles T. Menoher chief of the U.S. Army Air Service. Menoher had commanded the AEF's VI Corps.[45] The army demoted and court martialed Mitchell in 1925 after he refused to stop promoting his ideas about replacing other arms with airpower and winning wars entirely through massive strategic bombardments.

General Mason Patrick succeeded General Menoher in 1921. Though not as radical as Mitchell, General Patrick sought to expand the air service's capabilities. He established the Third Attack Group, dedicated to close air support, at Kelly Field, Texas. The unit's DeHavilland-4s, left over from the Great War, were not designed for flying ground support missions. Nevertheless, General Pershing refused to afford the air service the budget necessary to modernize its aircraft.[46]

Field Service Regulations 1923 was the first American tactical publication to describe aircraft as a combined arm. The manual placed aircraft in support of ground forces by providing observation and air cover, but made no mention of the long-range strategic bombing General Mitchell prescribed. The U.S. Army Air Service Tactical School at Langley Field, Virginia, further clarified air service doctrine in 1924. The school identified two distinct tasks for combat aircraft, "tactical missions" and "strategic missions." "Tactical missions" involved observation and close air support for ground forces. "Strategic missions" were aerial combat and long-range bombardment. General Patrick used his own terminology for these duties, calling ground support missions "air service" and air combat and bombing "air force."[47]

President Calvin Coolidge appointed Dwight W. Morrow head of a committee to assess American civilian and military aviation in 1925. Morrow had served as General Pershing's chief civilian advisor in France during the Great War. The Morrow Board's findings resulted in the Air Corps Act of 1926. This legislation transformed the U.S. Army Air Service into the U.S. Army Air Corps and outlined a five-year modernization plan. The air corps created two squadrons dedicated entirely to ground support. These squadrons conducted combined arms training exercises with the Second Infantry Division near San Antonio, Texas in 1927.[48]

Thanks to the Air Corps Act of 1926, the U.S. Army Air Corps possessed all-metal monoplanes by 1930. The air corps issued specifications for its first light bomber, designed specifically for close air support, in 1937. The Douglas A-20 Havoc had twin engines, a crew of four, a two thousand-pound bomb load, and seven .303-caliber

Conclusion

machine guns. Unfortunately, the air corps' wireless radio technology lagged. Dropping messages from the air was still the most reliable way for pilots to deliver information to infantry and artillery units in 1941.[49]

Despite the air corps' growth in the 1930s, the army remained unsure of the necessity of airpower in combined arms. The conflicts of the decade, such as the Italo-Ethiopian War, the Spanish Civil War, and the Second Sino-Japanese War, did little to convince the U.S. Army of the importance of aircraft. Most American commanders attributed the success of Italian, Fascist-Spanish, and Japanese airpower to the weakness of their Ethiopian, Republican-Spanish, and Chinese counterparts. Meanwhile, the *Luftwaffe* was preparing to play a major role in supporting the German Army's invasion of Poland in 1939.[50]

The U.S. Army's *Field Service Regulations 1939* reflected very few changes regarding aircraft as a combined arm since *Field Service Regulations 1923*. The 1939 publication referred to close air support as "light bombardment," but contained no specific tactical guidelines for its execution. The army did not clarify its close air support tactics until after the German *blitzkrieg* swept across Poland and France. The U.S. Army's September 1940 *Air Corps Memorandum* described aircraft providing covering fire for advancing ground forces. That same month, Assistant Chief of Staff for Operations Brigadier General Frank M. Andrews made a public statement criticizing the U.S. Army Air Corps for failing to develop air/ground combined arms at the same rate as its European counterparts. General Andrews had organized and supervised stateside aviation training during the Great War. He arranged for large air/ground training maneuvers, including one armored division, one motorized division, and one aerial bombardment wing, to begin at Fort Benning in winter 1940.[51] Nevertheless, the United States entered the Second World War before the air corps had the opportunity to perfect its ground-support tactics. The deadly skies above North Africa and the Southwest Pacific would be the U.S. Army Air Corps' training grounds. As for realizing William Mitchell's vision, the formation of a fully independent U.S. Air Force did not come to pass until 1947.

U.S. Army Combined Arms and Open Warfare After the Great War

Interwar Open Warfare Doctrine and Training

It was not a forgone conclusion that the U.S. Army would adopt General Pershing's brand of open warfare after the Great War. The six-week Meuse-Argonne Offensive was too short to provide a comprehensive showcase of AEF open warfare tactics. Some of Pershing's peers, such as former AEF Chief of Operations Major General Fox Conner, argued that the next war would see a return to trench warfare.[52] These mixed opinions on the nature of future conflicts resulted in the inconsistent development of Pershing's open warfare tactics in the 1920s and 1930s.

The U.S. Army's 1921 *Principles of War* and the *War Department Training Regulations Number 10–15* mentioned flexible formations, fire-and-maneuver, and battlefield improvisation. At the same time, these publications promoted the antiquated concept of *élan*. *Field Service Regulations 1923* detailed the execution of fire-and-maneuver and the importance of flanking enemy strong points, but also predicted that the next war would be one of static trench warfare.[53] U.S. Army training facilities, such as the Command and General Staff College at Fort Leavenworth, incorporated Pershing's open warfare concept into their curricula after the Great War, but budget cuts restricted the size and frequency of open warfare training exercises. The army could not afford to transport, feed, or house anything larger than a battalion in the 1920s. By 1930, there wasn't even enough money to mass produce U.S. Army training manuals. As a result, army training during the interwar period resembled pre–1917 garrison training, with marching drill, calisthenics, and target practice. Very few trainees participated in simulated combat maneuvers.[54] The U.S. Army Infantry School under George Marshall was an exception. Colonel Marshall worked hard to make Pershing's open warfare a priority when he became assistant commandant of the school in 1927. He and the eighty instructors under his supervision believed that teaching battlefield improvisation was key to turning out combat leaders who could maximize gains and minimize losses. Marshall criticized officers in the AEF for rarely planning beyond their first strategic objective and failing to improvise. As a result, most AEF units did not seize opportunities to capture ground

Conclusion

beyond their stated objective during the Meuse-Argonne Offensive. Marshall sought to remedy this by forcing his trainees to solve tactical problems through improvisation. Buses carried his students to unfamiliar training grounds without maps, forcing them to issue simple tactical commands that could be adjusted on the fly.[55]

Marshall criticized General Pershing's memoirs in 1931 for not acknowledging the strides in open warfare the AEF had made by the final phase of the Meuse-Argonne Offensive. Marshall's 1934 *Infantry in Battle* provided examples of the AEF's employment of infiltration tactics, fire-and-maneuver, and battlefield improvisation in 1918 as the blueprint for U.S. Army infantry tactics in the 1930s. He described infiltration as first line attackers bypassing strong points and exploiting weaknesses in the enemy line. The manual recommended flank attacks when neutralizing these enemy strong points. *Infantry in Battle* outlined a brand of fire-and-maneuver similar to Pershing's concept from the Great War. It claimed that direct suppressing fire was necessary for the survival and success of any attack. Colonel Marshall considered the portable automatic rifle to be the ideal infantry support weapon. *Infantry in Battle* placed the burden of battlefield survival and success on officers' ability to improvise. The book asserted that training needed to prepare officers to deal with imperfect and unpredictable battlefield conditions, worst case scenarios. Finally, Marshall expressed the importance of blending open warfare with advanced weapons. Throughout the book, he urged trainees to utilize all weapons at their disposal. Marshall intended for soldiers to learn all of these open warfare techniques through proper and thorough training prior to entering combat, not through bloody trial and error in battle as the AEF had done.[56]

Unfortunately, the open warfare tactics taught by Colonel Marshall at the U.S. Army Infantry School in the 1930s were not representative of the entire service during the interwar period. *Field Service Regulations 1939* mentioned nothing about General Pershing's open warfare that had not already been published in the 1923 regulations. The 1939 manual still asserted that *élan* could somehow carry soldiers across the battlefield. Chief of Staff of General Headquarters Lesley McNair openly criticized the infantry's assault tactics during the 1941 training operation in South Carolina as "faulty" and "deficient."[57]

U.S. Army Combined Arms and Open Warfare After the Great War

Meanwhile, rapid movement, surprise, flexible formations, and improvisation were already standard features in the tactical doctrine of the German Army that the Americans would face in battle a year later.

Conclusion

The U.S. Army transformed from a force suited only for regional limited warfare to an instrument of modern, industrial, global conflict from 1917 to 1918. The army, however, was only able to build upon the tactical lessons of the Great War to a limited extent during the interwar period. Combined arms tactics and General Pershing's open warfare concept remained a part of U.S. Army doctrine in the 1920s and 1930s, but the Great Depression and the nation's return to diplomatic isolationism interfered with the technological development and service wide training necessary to sustain this kind of advanced tactical growth. At the dawn of the Second World War, American soldiers still relied on artillery, machine guns, and mortars left over from the Great War. The GIs of 1941 even carried the same Springfield 1903 rifles as the doughboys. Most arms of the service were not yet motorized and wireless communication was not universal in 1941, creating the same overextended supply and communication lines and infantry/artillery coordination problems of 1918. The U.S. Army's version of Pershing's open warfare was a mix of modern concepts, such as flexible formations and improvisation, as well as antiquated notions, like *élan*. Furthermore, large-scale combined arms and open warfare training exercises were rare. Instead, marching drill and calisthenics dominated army training between the world wars. Needles to say, the U.S. Army returned to an experiential learning model during the Second World War. Undertrained, inexperienced, and ill-equipped for modern industrialized warfare, the GIs improvised their own brands of combined arms and open warfare on the battlefields of North Africa and the Southwest Pacific. Their advantage over the doughboys was the experienced leadership of generals MacArthur, Marshall, Patton, McNair, and Stilwell, who had survived the tactical learning curve of the western front twenty-three years earlier.

Conclusion

Legacy of the AEF

Ultimately, the AEF's way of war was and is consistent with the U.S. Army's tradition of learning through combat experience, rather than relying on peacetime tactical predictions of and preparations for the next conflict. This modality is apparent in the Global War on Terror of the twenty-first century. The U.S. Army spent the last three decades of the twentieth century preparing for a conventional war of attrition in Central Europe against the Soviet Red Army. In 1976, Commander of the U.S. Army Training and Doctrine Command General William E. Depuy wrote a series of manuals that became the army's doctrine and training guide for this anticipated conflict. General Depuy envisioned large-scale operations with combined arms coordination between mass armored/mechanized forces on the ground and airpower. The army clung to this doctrine even after the Soviet Union collapsed in 1991. The fall of the Soviet Union, however, brought military budget cuts and downsizing in the United States. This left the U.S. Army understrength, underequipped, and tactically unprepared for the challenges of the unconventional counter-insurgent warfare it would face in Afghanistan and Iraq in the early 2000s.[58]

In the first decade of the twenty-first century, the army still inclined toward combat experience and survival as its primary means of learning. America's hastily-trained National Guardsmen and volunteers were expected to learn while fighting and surviving in the deserts and cities of Iraq and the villages and mountains of Afghanistan. Trainees focused heavily on marksmanship and skipped entire segments of intended instruction to fulfill manpower needs overseas. In the early 2000s, the U.S. Army conducted few live-fire exercises due to the limited availability of weapons and equipment for instructional purposes just as the AEF did nearly one hundred years earlier. It was not uncommon for American artillerymen to fire their first live rounds in combat. Like the AEF, the U.S. Army sent new arrivals to "quiet areas" in Iraq and Afghanistan to gain experience through several weeks of raiding and patrolling. Furthermore, the army removed experienced troops from the combat zone for advanced training and returned them to the front to instruct their peers in the same way General Pershing did in 1918.

U.S. Army Combined Arms and Open Warfare After the Great War

In contrast with the AEF, the U.S. Army of the early twenty-first century enjoyed an abundance of updated instructional manuals and standardized training across the service. The army's basic training regimen included a degree of combined arms. Elementary infantry instruction introduced recruits to a variety of small arms, including light machine guns and anti-tank weapons. The army also offered advanced training to personnel designated to operate these special weapons in combat. The infantry trained alongside its support vehicles, such as personnel carriers and armored fighting vehicles. The U.S. Army of the early 2000s was similar to the AEF, however, in that it did not engage in large-scale combined arms training exercises involving infantry, artillery, armor, and aircraft together. The various sections of the army experienced tactical growth independently and at different paces.

The U.S. Army's most profound combined arms achievement of the early twenty-first century was reflected in its infantry squads and their tactics. A U.S. Army infantry squad in Iraq and Afghanistan was a mobile, mechanized, combined arms fighting unit in and of itself. Each squad of five to ten soldiers employed a variety of small arms, special weapons, and equipment, such as light machine guns, anti-tank guns, grenade launchers, and communications technology. Furthermore, the army provided its infantry squads with their own mobility, firepower, and protection by assigning each one an armored fighting vehicle. In this way, the U.S. Army of the early 2000s finally realized the post–World War I combined arms visions of Eisenhower, Marshall, and Patton.[59]

U.S. Army tactics and instruction in the early twenty-first century also retained elements of General Pershing's open warfare concept, such as fire-and-maneuver, the use of cover, and battlefield initiative and improvisation. Most American soldiers in Iraq and Afghanistan referred to fire-and-maneuver as "shoot-move-communicate." "Shoot-move-communicate" was part of the army's basic training program. Basic training also taught recruits to utilize cover as they advanced. The army continued to encourage its officers and noncommissioned officers to seize the initiative and improvise to survive and succeed in combat just as Pershing did in 1918.

Conclusion

The small amount of tactical training American soldiers received prior to their deployment hardly prepared them for the unconventional counter-insurgent operations they conducted in the Global War on Terror of the early twenty-first century. Nevertheless, they utilized innovative tactics in combat whenever applicable. The army embedded artillery observers within infantry companies to facilitate effective combined arms in combat, but infantry liaison with armor and air forces happened on a much more informal *ad hoc* basis. Rivalries between these arms also contributed to a lack of cooperation, as each sought to prove their ability to win battles singlehanded. Similar to the AEF in 1918, the U.S. Army of the early 2000s filled the infantry's manpower demands by cannibalizing existing artillery and armor units. This practice stripped the artillery and armor of experienced personnel and weakened the infantry with soldiers assigned to tasks they were not trained to perform.[60] Like the doughboys of the Great War and the GIs of the Second World War, American soldiers in the early years of the Global War on Terror learned to fight and survive in combat. Rather than attempt to predict and prepare for the tactical challenges of future conflicts, a goal seldom achieved in military history, the U.S. Army of the early twenty-first century continued its long-standing tradition of integrating new methods of warfare as it wages the next war.

Chapter Notes

Chapter I

1. Bruce I. Gudmundsson, *On Artillery* (London: Praeger, 1993), 46, 60, 70, 89; French General Staff, *French Trench Warfare, 1917–1918: A Reference Manual* (1918; repr., Nashville: Battery Press, 2002), 149, 316, 225.

2. Arthur L. Wagner, *Organization and Tactics: 8th Edition, 1918* (Chaumont, France: Headquarters of the U.S. Army, 1918), 43, 53, 154, 357, 360.

3. U.S. War Department, *Infantry Drill Regulations United States Army 1911, Corrected to April 15, 1917* (Washington, D.C.: Government Printing Office, 1917), 108.

4. U.S. War Department, *Field Artillery Notes No. 5* (Washington, D.C.: Government Printing Office, 1917), 10.

5. U.S. War Department, *Manual for Noncommissioned Officers and Privates of Field Artillery of the Army of the United States, Volume I, Corrected to December 31, 1917* (Washington, D.C.: Government Printing Office, 1918), 73, 20, 275.

6. U.S. War Department, *Manual for Noncommissioned Officers and Privates of Field Artillery of the Army of the United States, Volume II, Corrected to December 31, 1917* (Washington, D.C.: Government Printing Office, 1918), 8, 40.

7. U.S. War Department, *Instructions for the Offensive Combat of Small Units* (Washington, D.C.: Government Printing Office, 1918), 11, 27.

8. John J. Pershing, *Tactical Note Number 7, Combat Instructions for Troops of First Army* (1918), 1.

9. U.S. War Department, *Drill Regulations for the 75 French Gun Model 1897, Volume III* (Washington, D.C.: Government Printing Office, 1918), 68.

10. Frank J. Barber, *History of the Seventy-Ninth Division, A.E.F. During the World War: 1917–1919* (Lancaster, PA: Steinman and Steinman, 1922), 293.

11. John J. Pershing, *Combat Instructions* (General Headquarters: American Expeditionary Forces, 1918), 4, 7.

12. Conrad H. Lanza, "The Battle of the Meuse River: A River Crossing," *The Field Artillery Journal* 25, no. 5 (September–October 1935): 416.

13. American Expeditionary Forces General Headquarters, *Gas Manual, Part I: Tactical Employment of Gases* (1919), 19.

Chapter Notes—I

14. *Ibid.*, 22; Benedict Crowell and Robert F. Wilson, *Demobilization: Our Industrial and Military Demobilization after the Armistice: 1918–1920* (New Haven: Yale University Press, 1921), 223.

15. American Expeditionary Forces, *Tactical Employment of Gases*, 25.

16. American Expeditionary Forces General Headquarters, *Gas Manual, Part II: Use of Gas by the Artillery* (1919), 12.

17. American Expeditionary Forces General Headquarters, *Gas Manual, Part III: Use of Gas by Gas Troops* (1919), 116.

18. American Expeditionary Forces General Headquarters, *Gas Manual, Part IV: Use of Gas by Infantry* (1919), 9, 13, 18, 23.

19. Frederic L. Huidekoper, *The Military Unpreparedness of the United States: A History of American Land Forces from Colonial Times Until June 1, 1915* (New York: Macmillan, 1915), 477.

20. French General Staff, *French Trench Warfare*, 155; David A. Armstrong, *Bullets and Bureaucrats: The Machine Gun and the United States Army, 1861–1916* (Westport, CT: Greenwood Press, 1982), 154, 172, 192.

21. U.S. War Department, *Provisional Machine-Gun Firing Manual, 1917* (Washington, D.C.: Government Printing Office, 1917), 134, 190, 195.

22. U.S. War Department, *Notes on the Employment of Machine Guns* (Washington, D.C.: Government Printing Office, 1918), 10, 14, 21, 42, 46.

23. U.S. War Department, *Instructions for Small Units*, 13, 17; American Expeditionary Forces General Headquarters, *Employment of Machine Guns* (1918), 2.

24. Pershing, *Tactical Note Number 7*, 1; Pershing, *Combat Instructions*, 5.

25. U.S. War Department, *Instructions for Small Units*, 14, 33; Pershing, *Tactical Note Number 7*, 1, 3, 7; Pershing, *Combat Instructions*, 6, 9.

26. French General Staff, *French Trench Warfare*, 20, 165, 328.

27. U.S. War Department, *Notes on Grenade Warfare: Compiled from Data Available on February 15, 1917, Army War College* (Washington, D.C.: Government Printing Office, 1917), 7, 26; French General Staff, *French Trench Warfare*, 24.

28. U.S. War Department, *Grenade Warfare*, 35, 41.

29. U.S. War Department, *Instructions for the Training of Platoons for Offensive Action, 1917* (Washington, D.C.: Government Printing Office, 1917), 10; U.S. War Department, *Instructions for Small Units*, 12, 22; Pershing, *Combat Instructions*, 4.

30. French General Staff, *French Trench Warfare*, 158, 328.

31. U.S. War Department, *Training of Platoons*, 12, 25; U.S. War Department, *Employment of Machine Guns*, 13, 20.

32. U.S. War Department, *Manual of the Automatic Rifle (Chauchat), Drill-Combat-Mechanism* (Washington, D.C.: Government Printing Office, 1918), 17, 30.

33. *Ibid.*, 27, 34, 19.

34. U.S. War Department, *Instructions for Small Units*, 13, 22; Pershing, *Combat Instructions*, 4.

35. U.S. War Department, *Training of Platoons*, 10, 12; U.S. War Department, *Manual for Noncommissioned Officers and Privates of Infantry of the Army of the United States, 1917* (Washington, D.C.: Government Printing Office, 1917), 148.

36. U.S. War Department, *Small Arms Firing Manual 1913, Corrected to March 15, 1918* (Washington, D.C.: Government Printing Office, 1918), 134; French General Staff, *French Trench Warfare*, 18, 368.

37. U.S. War Department, *Instructions for Small Units*, 13, 14, 32.

38. Pershing, *Combat Instructions*, 6.

39. Frederic L. Huidekoper, *Illinois in the World War, Volume I: The History of the 33rd Division, A.E.F.* (Chicago: Illinois State Historical Library, 1921), 222.

40. David E. Johnson, *Fast Tanks and Heavy Bombers: Innovation in the U.S.*

Army, 1917–1945 (Ithaca: Cornell University Press, 1998), 31, 35.

41. Wagner, *Organization and Tactics*, 56; U.S. War Department, *Instructions for Small Units*, 11, 14, 33; Pershing, *Tactical Note Number 7*, 6.

42. Johnson, *Fast Tanks*, 42; Huidekoper, *Military Unpreparedness*, 472.

43. Johnson, *Fast Tanks*, 43, 47, 49.

44. Huidekoper, *Military Unpreparedness*, 471; Pershing, *Tactical Note Number 7*, 2, 6.

45. Barber, *History of the Seventy-Ninth Division*, 309.

46. Huidekoper, *Military Unpreparedness*, 471; U.S. War Department, *Instructions for Small Units*, 20.

Chapter II

1. U.S. War Department, *Manual for Noncommissioned Officers and Privates of Infantry*, 148.

2. Wagner, *Organization and Tactics*, 63, 98, 274, 282; Perry D. Jamieson, *Crossing the Deadly Ground: United States Army Tactics, 1865–1899* (Tuscaloosa, AL: University of Alabama Press, 1994), 43, 63, 71, 73, 105; Huidekoper, *Military Unpreparedness*, 202, 206, 209.

3. French General Staff, *French Trench Warfare*, 150, 33, 325.

4. Paddy Griffith, *Battle Tactics of the Western Front: The British Army's Art of Attack, 1916–18* (New Haven: Yale University Press, 1994), 96.

5. U.S. War Department, *Infantry Drill Regulations*, 9, 61, 84, 112, 116.

6. U.S. War Department, *Instructions for Small Units*, 16, 21; Pershing, *Tactical Note Number 7*, 3, 7; Pershing, *Combat Instructions*, 3.

7. Huidekoper, *Illinois in the World War*, 222.

8. French General Staff, *French Trench Warfare*, 33, 28.

9. U.S. War Department, *Infantry Drill Regulations*, 63,115.

10. *Ibid.*, 67, 104; U.S. War Department, *Manual for Noncommissioned Officers and Privates of Infantry*, 148; French General Staff, *French Trench Warfare*, 151.

11. U.S. War Department, *Infantry Drill Regulations*, 84, 105, 113; U.S. War Department, *Manual for Noncommissioned Officers and Privates of Infantry*, 150.

12. U.S. War Department, *Training of Platoons*, 12; U.S. War Department, *Manual for Noncommissioned Officers and Privates of Infantry*, 151.

13. Pershing, *Tactical Note Number 7*, 4; Pershing, *Combat Instructions*, 3.

14. French General Staff, *French Trench Warfare*, 33, 327, 153, 338.

15. Bruce I. Gudmundsson, *Stormtroop Tactics: Innovation in the German Army, 1914–1918* (London: Praeger, 1989), 66, 114; U.S. War Department, Document Number 802, *Instructions for Small Units*, 22, 35.

16. Pershing, *Tactical Note Number 7*, 3, 5; Pershing, *Combat Instructions*, 4, 6.

17. Huidekoper, *Illinois in the World War*, 215, 222.

18. Stephen E. Ambrose, *Upton and the Army* (Baton Rouge: Louisiana State University Press, 1964), 85; Jamieson, *Crossing Deadly Ground*, 18.

19. U.S. War Department, *Infantry Drill Regulations*, 63, 69, 82, 96, 116; U.S. War Department, *Manual for Noncommissioned Officers and Privates of Infantry*, 152.

20. U.S. War Department, *Training of Platoons*, 12, 14; U.S. War Department, *Instructions for Small Units*, 18, 28.

21. Pershing, *Tactical Note Number 7*, 1, 4; Pershing, *Combat Instructions*, 3; French General Staff, *French Trench Warfare*, 145, 33, 320, 338.

22. U.S. War Department, *Grenade Warfare*, 27, 30, 35.

23. U.S. War Department, *Training of Platoons*, 12; U.S. War Department, *Automatic Rifle*, 18, 28, 36.

Chapter Notes—III

24. U.S. War Department, *Instructions for Small Units*, 13; U.S. War Department, *Employment of Machine Guns*, 48; American Expeditionary Forces, *Employment of Machine Guns*, 2; U.S. War Department, *Employment of Machine Guns*, 48.

25. Pershing, *Tactical Note Number 7*, 3, 6; Pershing, *Combat Instructions*, 4; American Expeditionary Forces, *Gas Manual, Part II*, 14.

Chapter III

1. John F. O'Ryan, *The Story of the 27th Division* (New York: Wynkoop, Hallenbeck, Crawford Co., 1921), 143.
2. Huidekoper, *Military Unpreparedness*, 211, 271, 466.
3. General Staff, *Report on Foreign Maneuvers in 1912* (London: British War Office, 1913), 128, 20, 28, 49.
4. Huidekoper, *Illinois in the World War*, 3, 5, 15.
5. O'Ryan, *The Story of the 27th Division*, 128, 134.
6. *Ibid.*, 131.
7. Divisional Officers, *Official History of 82nd Division, American Expeditionary Forces: "All American" Division, 1917–1919* (Indianapolis: Bobbs-Merrill Co., 1919), 3.
8. Huidekoper, *Military Unpreparedness*, 212.
9. O'Ryan, *The Story of the 27th Division*, 141.
10. Huidekoper, *Military Unpreparedness*, 3.
11. Barber, *History of the Seventy-Ninth Division*, 28.
12. Divisional Officers, *Official History of 82nd Division*, 2.
13. John J. Pershing, "Training Cablegram Number 5," in *U.S. Army*, 14:318.
14. Harold B. Fiske, "Memorandum for the Chief of Staff," in *U.S. Army*, 14:303.
15. James G. Harbord, "Final Report of Assistant Chief of Staff, G-5," in *U.S. Army*, 14:290.
16. John J. Pershing, "General Orders Number 29: Instructions for Officers Visiting the French or British Lines or Serving with French, British, or American Units at the Front," in *U.S. Army*, 16:207.
17. John J. Pershing to Adjutant General, Cable No. 408-S, 22 December 1917, in *U.S. Army*, 14:319.
18. John J. Pershing to Adjutant General, Cable No. 952-S, 18 April 1918, in *U.S. Army*, 14:320; John J. Pershing to Chief of Staff, Cable No. 990-S, 24 April 1918, in *U.S. Army*, 14:322.
19. Peyton C. March to Commander-in-Chief, Cable No. 1259-R, 7 May 1918, in *U.S. Army*, 14:322; Peyton C. March to Commander-in-Chief, Cable No. 1543-R, 16 June 1918, in *U.S. Army*, 14:323; John J. Pershing to Secretary of War, Cable No. 1337-S, 19 June 1918, in *U.S. Army*, 14:324.
20. Army Dept., *U.S. Army*, 14:326; March, *The Nation at War*, 255, 294.
21. O'Ryan, *The Story of the 27th Division*, 119.
22. *Ibid.*, 121.
23. Paul B. Malone, "Memorandum for the Chief of Staff" in *U.S. Army*, 14:293.
24. James G. Harbord, "School Project for the American Expeditionary Forces," in *U.S. Army*, 14:293.
25. "I Corps Schools," Commander-in-Chief Report File, Folder No. 236, in *U.S. Army*, 14:397.
26. H. L. Cooper, "II Corps Schools," Commander-in-Chief Report File, Folder No. 237, in *U.S. Army*, 14:400.
27. "III Corps Schools," Commander-in-Chief Report File, Folder No. 238, in *U.S. Army*, 14:410.
28. Andre W. Brewster, "Final Report of the Inspector General, General Headquarters American Expeditionary Forces," Commander-in-Chief Report File, Folder No. 400, in *U.S. Army*, 15:303, 308, 312.
29. S. L. Pike, "Army Candidate School," Commander-in-Chief Report File, Folder No. 220, in *U.S. Army*, 14:340.
30. "Army School of the Line," Com-

Chapter Notes—III

mander-in-Chief Report File, Folder No. 219, in *U.S. Army*, 14:336.

31. "Army General Staff College," Commander-in-Chief Report File, Folder No. 218, in *U.S. Army*, 14:333; "Military Education in the American Expeditionary Forces," Commander-in-Chief Report File, Folder No. 248, in *U.S. Army*, 14:425.

32. John J. Pershing, "General Orders Number 35," in *U.S. Army*, 16:235.

33. John J. Pershing, "Bulletin Number 41," in *U.S. Army*, 17:73; John J. Pershing, "Bulletin Number 44," in *U.S. Army*, 17:75.

34. Army Dept., *U.S. Army*, 14:310, 298; Pershing, *My Experiences in the World War*, 391.

35. Huidekoper, *Illinois in the World War*, 6.

36. Ernest Hinds, "Final Report of the Chief of Artillery, American Expeditionary Forces," Commander-in-Chief Report File, Folder No. 381, in Army Dept., *U.S. Army in the World War*, 15:178; Army Dept., *U.S. Army*, 14:298; O'Ryan, *The Story of the 27th Division*, 138.

37. Adam F. Casad, "Appendix A [Ordnance]," Commander-in-Chief Report File, Folder No. 92, in Army Dept., *U.S. Army in the World War*, 14:82; Huidekoper, *Military Unpreparedness*, 473.

38. Leonard P. Ayres, *The War with Germany: A Statistical Summary* (Washington, D.C.: U.S. Government Printing Office, 1919), 80, 81, 94.

39. Hinds, "Final Report," 189, 196.

40. Fiske, "Memorandum for the Chief of Staff," 303; M.E. Lock, "Center of Artillery Studies Report," Commander-in-Chief Report File, Folder No. 221, in Army Dept., *U.S. Army in the World War*, 14:343.

41. U.S. Department of Army, Historical Division, *United States Army in World War II* (Washington, D.C.: U.S. Government Printing Office, 1950), 6:7; "Army Gas School," Commander-in-Chief Report File, Folder No. 228, in *U.S. Army*, 14:378; "Activities of the Chemical Warfare Service (July 5, 1917, to March 15, 1919)," Commander-in-Chief Report File, Folder No. 313, in *U.S. Army*, 15:292, 298.

42. O'Ryan, *The Story of the 27th Division*, 131; Divisional Officers, *Official History of 82nd Division*, 4.

43. John J. Pershing, "General Orders Number 82," in *U.S. Army*, 16:152; John J. Pershing, "Bulletin Number 30," in *U.S. Army*, 17:60; John J. Pershing, "General Orders Number 91," in *U.S. Army*, 16:341.

44. "Infantry Specialist School," 346, 350; French General Staff, *French Trench Warfare*, 159.

45. Johnson, *Fast Tanks*, 32.

46. George S. Patton, Jr., "Army Tank School Report," Commander-in-Chief Report File, Folder No. 229, in *U.S. Army*, 14:382.

47. Mason M. Patrick, "Final Report of Chief of Air Service, American Expeditionary Forces," Commander-in-Chief Report File, Folder No. 314, in *U.S. Army*, 15:236, 257, 262.

48. John J. Pershing, "General Orders Number 70," in *U.S. Army*, 16:315; John J. Pershing, "General Orders Number 81," in *U.S. Army*, 16:332.

49. Patrick, "Final Report," 233, 250, 280, 285.

50. John J. Pershing, "General Orders Number 77," in *U.S. Army*, 16:324; John J. Pershing, "General Orders Number 131," in *U.S. Army*, 16:411.

51. G. A. Youngberg, "Appendix C [Engineer]," Commander-in-Chief Report File, Folder 127, in *U.S. Army*, 14:148; George Van Horn Moseley, "Final Report of the Assistant Chief of Staff, G-4," Commander-in-Chief Report File, Folder 90, in *U.S. Army*, 14:77.

52. C. C. Carson, "Inspection of Motor Transportation," Commander-in-Chief Report File, Folder 149, in *U.S. Army*, 14:282.

53. Moseley, "Final Report," 77.

54. "Appendix F [Signal]," Commander-in-Chief Report File, Folder 169, in *U.S. Army*, 14:180; Brewster, "Final Report," 303, 308, 312.
55. O'Ryan, *The Story of the 27th Division*, 137.
56. Barber, *History of the Seventy-Ninth Division*, 35.
57. O'Ryan, *The Story of the 27th Division*, 170.
58. John J. Pershing to Benjamin Alvord, Jr., Cable No. P-952–5, 20 October 1917, in *U.S. Army*, 14:306.
59. U.S. War Department, "Training of Platoons," 15; "Infantry Specialist School," 350.
60. O'Ryan, *The Story of the 27th Division*, 246.
61. Liggett, *AEF Ten Years Ago*, 249, 252.
62. O'Ryan, *The Story of the 27th Division*, 245.
63. Blumenson, *Patton Papers*, 634.

Chapter IV

1. Hunter Liggett, "Use of Gas," G-3 Report File, I Corps, File 3004.04: Memo, in *U.S. Army*, 9:279; Hunter Liggett, "Employment of Gas," G-3 Report File, I Corps, File 3004.03: Memo, in *U.S. Army*, 9:306; George C. Marshall, "Operations Report Number 18," in *U.S. Army*, 9:311.
2. Pershing, *My Experiences*, 337, 350, 358.
3. Hunter Liggett, "Field Orders Number 88: First Army Prepares to Continue its Attack West of the Meuse," in *U.S. Army*, 9:335.
4. Hunter Liggett, "Battle Instructions of October 22, 1918, Annex Number 1: Employment of Army Artillery," in *U.S. Army*, 9:338.
5. Hunter Liggett, "Battle Instructions of October 22, 1918, Annex Number 2: Plan for Employment of Air Service Units, American First Army," in *U.S. Army*, 9:339.
6. Hunter Liggett, "Battle Instructions of October 22, 1918, Annex Number 3: Plan of Employment of Special Gas Troops," in *U.S. Army*, 9:340; Hunter Liggett, "Battle Instructions of October 22, 1918, Annex Number 5: Plan of Employment of Engineer Troops, Supply of Engineer Material, and Water Service," in *U.S. Army*, 9:343.
7. Hunter Liggett, "Battle Instructions of October 22, 1918, Annex Number 7: Means of Information, in U.S. Army," 9:348.
8. Marshall, *Memoirs*, 180; Liggett, *AEF Ten Years Ago*, 251.
9. Ferdinand Foch, "Foch's Views on Limited Objectives, Memorandum No. 5174," in *U.S. Army*, 8:103.
10. George C. Marshall, "Operations Report Number 25," in *U.S. Army*, 9:352; George C. Marshall, "Operations Report Number 27," in *U.S. Army*, 9:359.
11. Dennis E. Nolan, "Communiqués of the Assistant Chief of Staff, G-2, GHQ, AEF," in *U.S. Army*, 13:363; George C. Marshall, "Operations Report Number 28," in *U.S. Army*, 9:365.
12. Nolan, "Communiqués," 364.
13. U.S. Army War College, *Order of Battle of the United States Land Forces in the World War: American Expeditionary Forces* (1931; repr., Washington, D.C.: Center of Military History, United States Army, 1988), 1:133; U.S. Army War College, "Report of First Army," in *U.S. Army*, 9:367; Lanza, "Battle of the Meuse River," 393.
14. U.S. Army War College, "Report of First Army," in *U.S. Army*, 9:367.
15. Hunter Liggett, "Special Orders Number 518," in *U.S. Army*, 9:357; George C. Marshall, "Operations Report Number 29," in *U.S. Army*, 9:372.
16. Army War College, "Report of First Army," 367.
17. Liggett, *AEF Ten Years Ago*, 223; George C. Marshall, "Operations Report Number 32," in *U.S. Army*, 9:382.

Chapter Notes—IV

18. Nolan, "Communiqués," 366; Hunter Liggett, "Field Orders Number 102: First Army to Pursue Towards Raucourt," in *U.S. Army*, 9:384; George C. Marshall, "Operations Report Number 35," in *U.S. Army*, 9:394.

19. Nolan, "Communiqués," 368, 370, 374.

20. *Ibid.*, 369; Marshall, *Memoirs*, 192.

21. Pershing, *My Experiences*, 378; George C. Marshall, "Operations Report Number 34," in *U.S. Army*, 9:390; George C. Marshall, "Operations Report Number 36," in *U.S. Army*, 9:402; George C. Marshall, "Operations Report Number 37," in *U.S. Army*, 9:407.

22. Marshall, "Operations Report Number 37," 406–07; Marshall, *Memoirs*, 199; Lanza, "Battle of the Meuse River," 395.

23. Max von Gallwitz, "The Retreat to the Rhine," in *As They Saw Us*, ed. George S. Viereck (Garden City, NY: Doubleday, 1929), 282; Erich Ludendorff, *Ludendorff's Own Story, August 1914–November 1918: Volume II* (Harper and Brothers, 1919), 402.

24. Ludendorff, *Ludendorff's Own Story*, 420; Erich Ludendorff, "Order Number 10980," in *U.S. Army*, 9:563; Paul von Hindenburg, "Order Number 11213," in *U.S. Army*, 9:584.

25. Pershing, *My Experiences*, 371; Liggett, *AEF Ten Years Ago*, 215, 238; Georg von der Marwitz, "Order Number 545," in *U.S. Army*, 9:573; Marwitz, "Order Number 546," in *U.S. Army*, 9:573.

26. Huidekoper, *Illinois in the World War*, 204.

27. *Ibid.*, 197, 221.

28. *Ibid.*, 195, 200, 202.

29. Ludendorff, *Ludendorff's Own Story*, 398; Marshall, *Memoirs*, 175, 188; George C. Marshall, "Operations Report Number 38," in *U.S. Army*, 9:411.

30. U.S. Army War College Historical Section and Center of Military History, *Order of Battle of the United States Land Forces in the World War: American Expeditionary Forces* (1931; repr., Washington, D.C.: Center of Military History, United States Army, 1988), 2:79; Kenyon Stevenson, *The Official History of the Fifth Division USA: During the Period of its Organization and of its Operations in the European World War, 1917–1919* (Washington, D.C.: Society of the Fifth Division, 1919), 53.

31. Stevenson, *Official History of the Fifth Division*, 54; Lawrence L. Arbuckle to Clara Arbuckle Carroon, Ft. Leavenworth, KS, 18 February 1917, ed. Robert G. Carroon (2009).

32. Vernon G. Olsmith, *Recollections of an Old Soldier* (San Antonio, Texas: Vernon G. Olsmith, 1963), 20.

33. Stevenson, *Official History of the Fifth Division*, 57; American Battle Monuments Commission, *5th Division: Summary of Operations in the World War* (Washington, D.C.: U.S. Government Printing Office, 1944), 4, 6; Olsmith, *Recollections*, 22.

34. Stevenson, *Official History of the Fifth Division*, 59–61; John H. Smith, Personal Letter, 6 July 1918.

35. U.S. Army War College Historical Section and Center of Military History, *Order of Battle*, 2:81; American Battle Monuments Commission, *5th Division*, 8; Olsmith, *Recollections*, 31; Stevenson, *Official History of the Fifth Division*, 77, 84.

36. Harry M. Barthel, Notebook on Training, 22 September 1918, 23, 83, 88.

37. *Ibid.*, 86, 89.

38. American Battle Monuments Commission, *5th Division*, 15; Heldreth, *Company "D,"* 8; Lawrence L. Arbuckle to Clara Arbuckle Carroon, 30 September 1918, ed. Robert G. Carroon, 2009; Stevenson, *Official History of the Fifth Division*, 277.

39. Office of the Adjutant General of the U.S. Army, *American Decorations, 1862–1926* (Washington, D.C.: U.S. Government Printing Office, 1927), 120;

American Battle Monuments Commission, *5th Division*, 30, 33.

40. American Battle Monuments Commission, *5th Division*, 37.

41. Carl Noble, *Jugheads Behind the Lines* (Caldwell, ID: Caxton Printers, Ltd., 1938), 126.

42. Hanson E. Ely, "Field Orders Number 63: Orders for Next Push," in *Records of the World War: Field Orders 1918, 5th Division* (Washington, D.C.: U.S. Government Printing Office, 1921), 138; Hanson E. Ely, "Annex Number 1 to Field Orders Number 63: Plan for the Employment of the Air Service," in *Records of the World War*, 139.

43. Hanson E. Ely, "Operation Memorandum Number 119," in *Records of the World War*, 146; Stevenson, *Official History of the Fifth Division*, 303.

44. Stevenson, *Official History of the Fifth Division*, 187.

45. Stevenson, *Official History of the Fifth Division*, 304; Hanson E. Ely, "Field Orders Number 65," in *Records of the World War*, 152; American Battle Monuments Commission, *5th Division*, 42.

46. Hanson E. Ely, "Annex Number 2 to Field Orders Number 63," in *Records of the World War*, 141, 142, 144.

47. Stevenson, *Official History of the Fifth Division*, 191, 193, 198. American Battle Monuments Commission, *5th Division*, 42; Intelligence Section of the AEF General Staff, *Histories of Two Hundred and Fifty-One Divisions of the German Army Which Participated in the War (1914–1918)* (Chaumont, France: U.S. War Office, 1919), 125, 568.

48. Stevenson, *Official History of the Fifth Division*, 198, 201, 205; Lanza, "Battle of the Meuse River," 398; American Battle Monuments Commission, *5th Division*, 44; Hanson E. Ely, "Field Orders Number 68," in *Records of the World War*, 156.

49. Stevenson, *Official History of the Fifth Division*, 197, 207, 211, 213; American Battle Monuments Commission, *5th Division*, 46; Noble, *Jugheads*, 135.

50. Office of the Adjutant General of the U.S. Army, *American Decorations*, 2; American Battle Monuments Commission, *5thDivision*, 48; Stevenson, *Official History of the Fifth Division*, 213, 219, 221.

51. *Ibid.*, 217, 223; Hanson E. Ely, *Field Orders Number 71*, in *Official History of the Fifth Division*, 309.

52. Stevenson, *Official History of the Fifth Division*, 233; Noble, *Jugheads*, 146; Stevenson, *Official History of the Fifth Division*, 235.

53. Stevenson, *Official History of the Fifth Division*, 239, 241, 245, 310; John Cherpak, Battle Account, 6 November 1918; Clyde Heldreth, *History of Company "D," 60th Infantry, U.S. Army* (Esch, Luxemburg: Clyde Heldreth, 1919), 13.

54. Hanson E. Ely, "Secret Field Orders Number 73," in *Official History of the Fifth Division*, 311; Noble, *Jugheads*, 149.

55. Stevenson, *Official History of the Fifth Division*, 251; Paul A. Doty, Personal Letter, 26 November 1918; Ely, "Secret Field Orders Number 75," in *Records of the World War*, 165.

56. Lanza, "Battle of the Meuse River," 411.

Conclusion

1. William O. Odom, *After the Trenches: The Transformation of U.S. Army Doctrine, 1918–1939* (College Station: Texas A&M University Press, 1999), 7; Frank E. Vandiver, *Black Jack: The Life and Times of John J. Pershing* (College Station: Texas A&M University Press, 1977), 1060; Coffman, *The Regulars: The American Army, 1898–1941*, 226, 281; James L. Abrahamson, *American Arms for a New Century: The Making of a Great Military Power* (New York: Free Press, 1981), 178.

2. Brian M. Linn, *The Echo of Battle: The Army's Way of War* (Cambridge: Harvard University Press, 2007), 129; Russell F. Weigley, *The American Way of War: A History of United States Military Strategy*

Chapter Notes—Conclusion

and Policy (Bloomington: Indiana University Press, 1973), 213; Coffman, *The Regulars*, 263; Odom, *After the Trenches*, 237, 241, 49.

3. Odom, *After the Trenches*, 49, 204; Coffman, *The Regulars*, 264.

4. Peter Maslowski and Alan R. Millett, *For the Common Defense: A Military History of the United States of America* (New York: The Free Press, 1984), 380; Johnson, *Fast Tanks*, 113; U.S. Department of Army, Historical Division, *United States Army in World War II* (Washington, D.C.: U.S. Government Printing Office, 1950), 6: 25.

5. House, *Combined Arms Warfare*, 65; Linn, *Echo of Battle*, 116.

6. Johnson, *Fast Tanks*, 56, 67; Odom, *After the Trenches*, 92.

7. Odom, *After the Trenches*, 240, 142; House, *Combined Arms Warfare*, 97; U.S. Department of Army, Historical Division, *United States Army in World War II* (Washington, D.C.: U.S. Government Printing Office, 1950), 1: 260.

8. Lanza, "Battle of the Meuse River," 393.

9. Infantry Journal, Inc., *Infantry in Battle* (Military History and Publication Section of the Infantry School, 1934), 223, 273, 250.

10. Bland and Ritenour, *Papers of George Catlett Marshall*, 373; Infantry Journal, *Infantry in Battle*, 274; Forrest C. Pogue, *George C. Marshall: Education of a General, 1880–1939* (New York: Viking Press, 1963), 253.

11. Odom, *After the Trenches*, 157, 139; House, *Combined Arms Warfare*, 101.

12. Odom, *After the Trenches*, 166, 158.

13. Crowell and Wilson, *Demobilization*, 164, 174.

14. Janice McKenney, "More Bang for the Buck in the Interwar Army: The 105mm Howitzer," *Military Affairs* 42 (April 1978): 81; Odom, *After the Trenches*, 154.

15. McKenney, "Interwar Army," 81; Odom, *After the Trenches*, 62, 156.

16. Coffman, *The Regulars*, 231; Crowell and Wilson, *Demobilization*, 220; Odom, *After the Trenches*, 69; U.S. Department of Army, Historical Division, *United States Army in World War II*, 6:15.

17. Crowell and Wilson, *After the Trenches*, 221, 226; U.S. Department of Army, Historical Division, *United States Army in World War II*, 6:200.

18. Infantry Journal, *Infantry in Battle*, 247, 223, 239; Odom, *After the Trenches*, 141, 183, 180, 139; Crowell and Wilson, *Demobilization*, 196.

19. Gudmundsson, *On Artillery*, 110, 112; Infantry Journal, *Infantry in Battle*, 224; Odom, *After the Trenches*, 58, 141.

20. Odom, *After the Trenches*, 174, 183; Infantry Journal, *Infantry in Battle*, 223.

21. Odom, *After the Trenches*, 43, 46; Walter Millis, *Arms and Men: A Study in American Military History* (New York: Mentor Books, 1956), 215.

22. Odom, *After the Trenches*, 54; Linn, *Echo of Battle*, 120, 133; Johnson, *Fast Tanks*, 34.

23. Coffman, *The Regulars*, 229; Carlo D'Este, *Patton: A Genius for War* (New York: HarperCollins, 1995), 285; Blumenson, *Patton Papers*, 701, 707, 715; Crowell and Wilson, *Demobilization*, 195.

24. Blumenson, *Patton Papers*, 716, 720, 735; D'Este, *Patton: A Genius*, 297.

25. Dwight D. Eisenhower, "A Tank Discussion," *Infantry Journal* 17 (November 1920): 453, 454, 456.

26. Blumenson, *Patton Papers*, 717, 719, 721, 728.

27. Johnson, *Fast Tanks*, 57, 72, 75, 77; Odom, *After the Trenches*, 57, 90; Maslowski and Millett, *Common Defense*, 381.

28. Johnson, *Fast Tanks*, 32, 68, 73, 77; Weigley, *History of the United States Army*, 409.

29. Johnson, *Fast Tanks*, 73, 77, 80.

30. Odom, *After the Trenches*, 178, 91; Coffman, *The Regulars*, 268; Johnson, *Fast Tanks*, 96; Weigley, *History of the United States Army*, 410.

Chapter Notes—Conclusion

31. Johnson, *Fast Tanks*, 99; Blumenson, *Patton Papers*, 881, 884; Odom, *After the Trenches*, 92.

32. Maslowski and Millett, *Common Defense*, 382; Johnson, *Fast Tanks*, 117.

33. Johnson, *Fast Tanks*, 118, 120, 121; Odom, *After the Trenches*, 96.

34. Odom, *After the Trenches*, 142, 184.

35. Johnson, *Fast Tanks*, 141; Odom, *After the Trenches*, 145, 242; U.S. Department of Army, Historical Division, *United States Army in World War II*, 6: 210, 237; U.S. Department of Army, Historical Division, *United States Army in World War II*, 1: 60.

36. House, *Combined Arms Warfare*, 67; Odom, *After the Trenches*, 63; March, *The Nation at War*, 275.

37. Crowell and Wilson, *Demobilization*, 230; Millis, *Arms and Men*, 215; Linn, *Echo of Battle*, 119, 135; Blumenson, *Patton Papers*, 702, 741, 841, 844; D'Este, *Patton: A Genius*, 302.

38. Bradford G. Chynoweth, "Cavalry Tanks," *Cavalry Journal* 30 (January 1921): 248.

39. Johnson, *Fast Tanks*, 74, 124; Blumenson, *Patton Papers*, 841, 842, 847, 850, 865; Odom, *After the Trenches*, 65.

40. Johnson, *Fast Tanks*, 116, 130; Weigley, *History of the United States Army*, 410; Odom, *After the Trenches*, 146; Coffman, *The Regulars*, 270; Blumenson, *Patton Papers*, 884.

41. Johnson, *Fast Tanks*, 131.

42. *Ibid.*, 135, 139; Coffman, *The Regulars*, 271; House, *Combined Arms Warfare*, 102; Odom, *After the Trenches*, 148.

43. Johnson, *Fast Tanks*, 57, 100; Linn, *Echo of Battle*, 132.

44. Crowell and Wilson, *Demobilization*, 205, 207, 210.

45. House, *Combined Arms Warfare*, 68; Johnson, *Fast Tanks*, 81; U.S. Department of Army, Historical Division, *United States Army in World War II*, 6: 283.

46. Johnson, *Fast Tanks*, 68, 85; Odom, *After the Trenches*, 66, 159; Vandiver, *Black Jack*, 1060; Linn, *Echo of Battle*, 126.

47. Odom, *After the Trenches*, 66, 89; Johnson, *Fast Tanks*, 92; U.S. Department of Army, Historical Division, *United States Army in World War II*, 6: 283.

48. House, *Combined Arms Warfare*, 103; Odom, *After the Trenches*, 89, 159.

49. Odom, *After the Trenches*, 90, 161.

50. *Ibid.*, 138, 160, 184.

51. *Ibid.*, 162, 242; House, *Combined Arms Warfare*, 103; U.S. Department of Army, Historical Division, *United States Army in World War II*, 1: 102, 104, 106, 110; U.S. Department of Army, Historical Division, *United States Army in World War II*, 6: 239.

52. Linn, *Echo of Battle*, 120; Russ Stayanoff, "Major General Fox Conner: Soldier, Mentor, Enigma: Operations Chief (G3) of the AEF," www.worldwar1.com/dbc/foxconner.htm; Odom, *After the Trenches*, 175.

53. Weigley, *American Way of War*, 213; Linn, *Echo of Battle*, 129; Johnson, *Fast Tanks*, 96; Odom, *After the Trenches*, 41, 241.

54. Coffman, *The Regulars*, 281; Johnson, *Fast Tanks*, 56; Odom, *After the Trenches*, 95.

55. Odom, *After the Trenches*, 87; Bland and Ritenour, *Papers of George Catlett Marshall*, 320, 323, 335; Pogue, *Education of a General*, 249, 251.

56. Bland and Ritenour, *Papers of George Catlett Marshall*, 338, 367; Infantry Journal, *Infantry in Battle*, VII, 55, 94, 223, 307, 322.

57. House, *Combined Arms Warfare*, 96; Odom, *After the Trenches*, 138, 185; U.S. Department of Army, Historical Division, *United States Army in World War II*, 6: 239; U.S. Department of Army, Historical Division, *United States Army in World War II*, 1: 33.

58. Linn, *Echo of Battle*, 204, 211, 222.

59. Marcus Peterson (Sergeant, U.S. Army National Guard) interview with the author, April 2010.

60. *Ibid.*

BIBLIOGRAPHY

Archival Materials

National Archives. College Park, MD

Pershing, John J. "Tactical Note Number 7, Combat Instructions for Troops of First Army, 29 August 1918." Record Group 120. National Archives.

U.S. Army Heritage and Education Center. Carlisle Barracks, PA

American Expeditionary Forces General Headquarters. "Bulletin Number 30, Employment of Machine Guns." 23 May 1918. Carlisle Barracks, PA: U.S. Army Heritage and Education Center.

_____. *Gas Manual, Part I: Tactical Employment of Gases.* March 1919. Carlisle Barracks, PA: U.S. Army Heritage and Education Center.

_____. *Gas Manual, Part II: Use of Gas by the Artillery.* March 1919. Carlisle Barracks, PA: U.S. Army Heritage and Education Center.

_____. *Gas Manual, Part III: Use of Gas by Gas Troops.* March 1919. Carlisle Barracks, PA: U.S. Army Heritage and Education Center.

_____. *Gas Manual, Part IV: Use of Gas by Infantry.* March 1919. Carlisle Barracks, PA: U.S. Army Heritage and Education Center.

Barthel, Harry M. Notebook on Training. WWI Veterans Survey Inventories. 22 September 1918. Carlisle Barracks, PA: U.S. Army Heritage and Education Center.

Cherpak, John. Battle Account. WWI Veterans Survey Inventories. 6 November 1918. Carlisle Barracks, PA: U.S. Army Heritage and Education Center.

Doty, Paul A. Personal Letter. WWI Veterans Survey Inventories. 26 November 1918. Carlisle Barracks, PA: U.S. Army Heritage and Education Center.

Pershing, John J. *Combat Instructions.* 5 September 1918. Carlisle Barracks, PA: U.S. Army Heritage and Education Center.

Bibliography

Smith, John H. Personal Letter. WWI Veterans Survey Inventories, 6 July 1918. Carlisle Barracks, PA: U.S. Army Heritage and Education Center.

United States War Department. *Drill Regulations for the 75 French Gun Model 1897, Volume III.* September 1918. Document Number 875. Carlisle Barracks, PA: U.S. Army Heritage and Education Center.

_____. *Field Artillery Notes No. 5.* 27 June 1917. Document Number 619. Carlisle Barracks, PA: U.S. Army Heritage and Education Center.

_____. *Infantry Drill Regulations United States Army 1911, Corrected to April 15, 1917.* 1917. Document Number 394. Carlisle Barracks, PA: U.S. Army Heritage and Education Center.

_____. *Instructions for the Offensive Combat of Small Units.* May 1918. Document Number 802. Carlisle Barracks, PA: U.S. Army Heritage and Education Center.

_____. *Instructions for the Training of Platoons for Offensive Action, 1917.* 14 June 1917. Document Number 613. Carlisle Barracks, PA: U.S. Army Heritage and Education Center.

_____. *Manual of the Automatic Rifle (Chauchat), Drill-Combat-Mechanism.* April 1918. Document Number 793. Carlisle Barracks, PA: U.S. Army Heritage and Education Center.

_____. *Manual for Noncommissioned Officers and Privates of Field Artillery of the Army of the United States, Volume I, Corrected to December 31, 1917.* 1918. Document Number 614. Carlisle Barracks, PA: U.S. Army Heritage and Education Center.

_____. *Manual for Noncommissioned Officers and Privates of Infantry of the Army of the United States, 1917.* 1917. Document Number 574. Carlisle Barracks, PA: U.S. Army Heritage and Education Center.

_____. *Notes on the Employment of Machine Guns.* 1918. Document Number 712. Carlisle Barracks, PA: U.S. Army Heritage and Education Center.

_____. *Notes on Grenade Warfare: Compiled from Data Available on February 15, 1917, Army War College.* 28 April 1917. Document Number 576. Carlisle Barracks, PA: U.S. Army Heritage and Education Center.

_____. *Provisional Machine-Gun Firing Manual, 1917.* 21 June 1917. Document Number 615. Carlisle Barracks, PA: U.S. Army Heritage and Education Center.

Letters

Arbuckle, Lawrence L., to Clara Arbuckle Carroon. Ft. Leavenworth, KS, 18 February 1917. Edited by Robert G. Carroon. 2009.

Books

American Battle Monuments Commission. *5th Division: Summary of Operations in the World War.* Washington, D.C.: U.S. Government Printing Office, 1944.

Bibliography

American Expeditionary Forces General Staff. Intelligence Section. *Histories of Two Hundred and Fifty-One Divisions of the German Army Which Participated in the War (1914–1918)*. Chaumont, France: U.S. War Office, 1919.

Ayres, Leonard P. *The War with Germany: A Statistical Summary*. Washington, D.C.: U.S. Government Printing Office, 1919.

Barber, J. Frank. *History of the Seventy-Ninth Division, A.E.F. During the World War: 1917–1919*. Lancaster, PA: Steinman and Steinman, 1922.

Blumenson, Martin, *The Patton Papers, 1885–1940*. Boston: Houghton Mifflin, 1972.

British War Office General Staff. *Report on Foreign Maneuvers in 1912*. London: British War Office, 1913.

Crowell, Benedict, and Robert F. Wilson, *Demobilization: Our Industrial and Military Demobilization after the Armistice: 1918–1920*. New Haven: Yale University Press, 1921.

French Army General Staff. *French Trench Warfare, 1917–1918: A Reference Manual*. Nashville: Battery Press, 2002. First published 1918 by the French Army.

Gallwitz, Max von. "The Retreat to the Rhine." In *As They Saw Us*, edited by George S. Viereck 230–87. Garden City, NY: Doubleday, 1929.

Heldreth, Clyde. *History of Company "D," 60th Infantry, U.S. Army*. Esch, Luxemburg: Clyde Heldreth, 1919.

Huidekoper, Frederic L. *Illinois in the World War, Volume I: The History of the 33rd Division, A.E.F.* Chicago: Illinois State Historical Library, 1921.

———. *The Military Unpreparedness of the United States: A History of American Land Forces from Colonial Times Until June 1, 1915*. New York: Macmillan, 1915.

Infantry Journal, Inc. *Infantry in Battle*. Washington, D.C.: Military History and Publication Section of the Infantry School, 1934.

Liggett, Hunter. *AEF Ten Years Ago in France*. New York: Dodd and Mead, 1928.

Ludendorff, Erich. *Ludendorff's Own Story, August 1914–November 1918: Volume II*. New York: Harper and Brothers, 1919.

March, General Peyton Conway. *The Nation at War*. Garden City, NY: Doubleday, Doran & Company, 1932.

Marshall, George C. *The Papers of George Catlett Marshall: Volume 1, "The Soldierly Spirit" December 1880–June 1939*. Edited by Larry I. Bland and Sharon R. Ritenour. Baltimore: Johns Hopkins University Press, 1981.

Noble, Carl. *Jugheads Behind the Lines*. Caldwell, ID: Caxton Printers, Ltd., 1938.

Official History of 82nd Division, American Expeditionary Forces: "All American" Division, 1917–1919 by Divisional Officers. Indianapolis: Bobbs-Merrill Co., 1919.

Olsmith, Vernon G. *Recollections of an Old Soldier*. San Antonio: The author, 1963.

O'Ryan, John F. *The Story of the 27th Division*. New York: Wynkoop, Hallenbeck, Crawford Co., 1921.

Bibliography

Pershing, John J. *My Experiences in the World War.* New York: Frederick A. Stokes Company, 1931.

Records of the World War: Field Orders 1918, 5th Division. Washington, D.C.: U.S. Government Printing Office, 1921.

Stevenson, Kenyon. *The Official History of the Fifth Division USA: During the Period of its Organization and of its Operations in the European World War, 1917–1919.* Washington, D.C.: Society of the Fifth Division, 1919.

United States. Department of the Army. Historical Division. *United States Army in World War II, Volume 6: The War Department, Chief of Staff: Prewar Plans and Preparations.* Washington, D.C.: Government Printing Office, 1950.

____. ____. ____. *United States Army in the World War, 1917–1919, Volume 8: Military Operations of the American Expeditionary Forces.* Washington, D.C.: U.S. Army Center of Military History, 1989. First published 1948 by the Government Printing Office.

____. ____. ____. *United States Army in the World War, 1917–1919, Volume 9: Military Operations of the American Expeditionary Forces.* Washington, D.C.: U.S. Army Center of Military History, 1989. First published 1948 by the Government Printing Office.

____. ____. ____. *United States Army in the World War, 1917–1919, Volume 13: Reports of the Commander-in-Chief, Staff Sections and Services.* Washington, D.C.: U.S. Army Center of Military History, 1989. First published 1948 by the Government Printing Office.

____. ____. ____. *United States Army in the World War, 1917–1919, Volume 14: Reports of the Commander-in-Chief, Staff Sections and Services.* Washington, D.C.: U.S. Army Center of Military History, 1989. First published 1948 by the Government Printing Office.

____. ____. ____. *United States Army in the World War, 1917–1919, Volume 15: Reports of the Commander-in-Chief, Staff Sections and Services.* Washington, D.C.: U.S. Army Center of Military History, 1989. First published 1948 by the Government Printing Office.

____. ____. ____. *United States Army in the World War, 1917–1919, Volume 16: General Orders, GHQ, AEF.* Washington, D.C.: U.S. Army Center of Military History, 1989. First published 1948 by the Government Printing Office.

____. ____. ____. *United States Army in the World War, 1917–1919, Volume 17: Bulletins, GHQ, AEF.* Washington, D.C.: U.S. Army Center of Military History, 1989. First published 1948 by the Government Printing Office.

____. ____. Army War College Historical Section and Center of Military History. *Order of Battle of the United States Land Forces in the World War: American Expeditionary Forces, Volume 1: General Headquarters, Armies, Army Corps, Services of Supply, Separate Forces.* Washington, D.C.: U.S. Army Center of Military History, 1988. First published 1931 by the Government Printing Office.

____. ____. *Order of Battle of the United States Land Forces in the World War:*

Bibliography

American Expeditionary Forces, Volume 2: Divisions. Washington, D.C.: U.S. Army Center of Military History, 1988. First published 1931 by The Government Printing Office.

Wagner, Arthur L. *Organization and Tactics: 8th Edition, 1918.* Chaumont, France, Headquarters of the Army, 1918.

Monographs

Abrahamson, James L. *American Arms for a New Century: The Making of a Great Military Power.* New York: Free Press, 1981.

Ambrose, Stephen E. *Upton and the Army.* Baton Rouge: Louisiana State University Press, 1964.

Coffman, Edward M. *The Regulars: The American Army, 1898–1941.* Cambridge: Harvard University Press, 2004.

D'Este, Carlo. *Patton: A Genius for War.* New York: HarperCollins, 1995.

Griffith, Paddy. *Battle Tactics of the Western Front: The British Army's Art of Attack, 1916–18.* New Haven: Yale University Press, 1994.

Gudmundsson, Bruce I. *On Artillery.* London: Praeger, 1993.

_____. *Stormtroop Tactics: Innovation in the German Army, 1914–1918.* London: Praeger, 1989.

House, Jonathan M. *Combined Arms Warfare in the Twentieth Century.* Lawrence: University Press of Kansas, 2001.

Jamieson, Perry D. *Crossing the Deadly Ground: United States Army Tactics, 1865–1899.* Tuscaloosa: University of Alabama Press, 1994.

Johnson, David E. *Fast Tanks and Heavy Bombers: Innovation in the U.S. Army, 1917–1945.* Ithaca: Cornell University Press, 1998.

Linn, Brian M. *The Echo of Battle: The Army's Way of War.* Cambridge: Harvard University Press, 2007.

Maslowski, Peter, and Alan Reed Millett. *For the Common Defense: A Military History of the United States of America.* New York: The Free Press, 1984.

Millis, Walter. *Arms and Men: A Study in American Military History.* New York: Mentor Books, 1956.

Odom, William O. *After the Trenches: The Transformation of U.S. Army Doctrine, 1918–1939.* College Station: Texas A&M University Press, 1999.

Pogue, Forrest C. *George C. Marshall: Education of a General, 1880–1939.* New York: Viking Press, 1963.

Vandiver, Frank Everson. *Black Jack: The Life and Times of John J. Pershing.* College Station: Texas A&M University Press, 1977.

Weigley, Russell F. *The American Way of War: A History of United States Military Strategy and Policy.* Bloomington: Indiana University Press, 1973.

_____. *History of the United States Army.* New York: Macmillan, 1967.

Bibliography

Articles

Chynoweth, Bradford G. "Cavalry Tanks." *Cavalry Journal* 30 (January 1921): 247–51.

Eisenhower, Dwight D. "A Tank Discussion." *Infantry Journal* 17 (November 1920): 453–58.

Lanza, Conrad H. "The Battle of the Meuse River: A River Crossing." *The Field Artillery Journal* 25 (September–October 1935): 393–416.

McKenney, Janice. "More Bang for the Buck in the Interwar Army: The 105mm Howitzer." *Military Affairs* 42 (April 1978): 81–85.

Stayanoff, Russ. "Major General Fox Conner: Soldier, Mentor, Enigma: Operations Chief (G3) of the AEF." Doughboy Center: The Story of the American Expeditionary Forces. Accessed 11 November 2010. www.worldwar1.com/dbc/fox conner.htm.

Dissertation

Grotelueschen, Mark Ethan. "The AEF Way of War: The American Army and Combat in the First World War." PhD diss., Texas A&M University, 2004.

INDEX

Aberdeen Proving Ground, Maryland 133
Activities of the Chemical Warfare Service (July 5, 1917, to March 15, 1919) Report 72
aerial bombardment 36, 37, 92, 94–97, 116, 130, 140–42
aerial bombardment training 77, 140–42
aerial reconnaissance 5, 36, 51, 56, 57, 71, 77, 89, 90, 92, 94, 95, 108–10, 117, 133, 135, 139–41
aerial reconnaissance training 77–79, 140–42
aerodrome 76, 92
Aeronautical Engine Plant, Long Island, New York 140
Afghanistan (War on Terror) 146, 147
Afrika Korps 137
Aincreville, France 95, 109
Air Corps (U.S.) 141, 142
Air Corps Act (1926) 141
Air Corps Memorandum 142
"air force" 141
Air Force (U.S.) 142
"air service" 141
Air Service (U.S.) 37, 71, 76–78, 80, 140, 141
air superiority 92, 95, 140
air support 5, 8, 35–37, 51, 78, 89, 90, 92, 94, 95, 104, 105, 140, 142
air support training 65, 67, 76–80, 140–42
aircraft 1, 4–8, 13, 19, 22, 35–37, 39, 40, 51, 56, 61, 87, 94–96, 118, 122–24, 139, 140, 147, 148
airfield 140
Allworth, Captain Edward 114
Alvord, General Benjamin, Jr. 62, 85
Ambulance Company Thirty (U.S.) 115
Andon River, France 109, 110
Andrews, General Frank M. 142
Annex Number 1 to Field Orders Number 63 (Ely) 108
Annex Number 2 to Field Orders Number 63 (Ely) 109, 110
Annual Report of the Chief of Staff (MacArthur) 124
Anould Sector, France 104
Arbuckle, Private Lawrence L. 102, 106
Argonne Forest, France 65
armistice (1918) 2, 3, 10, 27, 49, 54, 70, 82, 88, 89, 97, 98, 100, 118–22, 127, 130, 138, 140
armored car 135, 138, 139, 147
Armored Force (U.S.) 137
Army Gas School Report 72
Army Tank School Report (Patton) 75
artillery 1, 4, 6, 8, 13–19, 22, 23, 25, 29, 33, 35, 39, 40, 44, 55–57, 59, 61, 62, 68, 72, 75, 76, 78, 80, 82–84, 87, 88, 90–110, 113–18, 122–24, 127, 135, 139, 140, 142, 145–48
artillery factory 127
artillery training 56, 59, 62, 65, 67, 69–71, 101–03, 126–28
assault column (infantry) 41–44, 49, 52, 85

Index

assault troops 52
assault wave (infantry) 48–50, 53
attrition (strategy) 146
automatic rifle 4, 13, 30, 39, 40, 45–47, 51–53, 59, 61, 105, 109, 113, 117, 124, 125, 131, 144
automatic rifle training 65–67, 73, 74, 130–32
automobile 5, 6, 8, 13, 19, 25, 26, 28, 56, 69, 70, 78, 81, 82, 87, 93, 94, 96, 98, 99, 115, 116, 118, 124, 126, 128, 129, 135, 136, 138, 142, 145
Aviation Section of the Army Signal Corps (U.S.) 35
Ayres, Colonel Leonard P. 70

Baker, Secretary Newton D. 63
Balkan Wars (1912) 28
Bangalore torpedo 38
Bantheville, France 91, 107
banzai charge 32
barbed wire 18, 38, 91
Barricourt, France 94
Bar-sur-Aube, France 9, 102, 103
Barthel, Private Harry M. 104, 105
battery (artillery) 9, 15, 18–20, 22, 27
Battle Instructions of October 22, 1918 (Liggett) 90
Battle Instructions of October 22, 1918, Annex Number 1: Employment of Army Artillery (Liggett) 91
Battle Instructions of October 22, 1918, Annex Number 2: Plan for Employment of Air Service Units, American First Army (Liggett) 92
Battle Instructions of October 22, 1918, Annex Number 3: Plan if Employment of Special Gas Troops (Liggett) 92
Battle Instructions of October 22, 1918, Annex Number 5: Plan of Employment of Engineer Troops, Supply of Engineer Material, and Water Service (Liggett) 92
Battle Instructions of October 22, 1918, Annex Number 7: Means of Information (Liggett) 92
Baumont, France 92
bayonet 6, 30, 32–34, 40, 45–47, 51, 57–59, 61, 64
bayonet training 65, 66
Bealon, France 97
Belgium 100
Bjornstad, General Alfred W. 67
blitzkrieg 137, 142
Boer Wars (1899) 38, 43, 45

Bois de Babiemont, France 108–110
Bois de Bantheville, France 90
Bois de Barricourt, France 90
Bois de Bourgogne, France 90, 95
Bois de la Pultiere, France 106, 107
Bois de Sassey, France 90
Bois de Warville, France 100
Bois des Rappes, France 90, 106, 107
Bois d'Harville, France 49
Bois Hanido, France 106
Bonvaux, France 106
Bordeaux, France 93
Borne du Cornouiller, France 96
Bourg, France 75
Bovington, England 75
box barrage (artillery) 14
Brandeville, France 97, 117
Breguet 14 aircraft 36
Bren Gun 131
Brett, Major Sereno E. 135
Brewer, Major Carlos B. 127
Brewster, General Andre W. 66, 68, 83
bridge construction 80, 89, 97, 119
Brieulles, France 112, 116
British Army 2, 4, 6, 13, 15, 19, 20, 23, 24, 27–30, 32–34, 36, 38, 39, 43, 45, 47, 51, 52, 56–61, 63, 75, 79, 86, 87, 101, 130, 131, 135
bromobenzylcyanide gas 21, 22
Browning M1918 Automatic Rifle (BAR) 74, 131
Browning M1917 machine gun 25, 134
Browning M2 machine gun 131
Bruchmuller, Colonel Georg 15
Bulletin Number 30 (Pershing) 73
Bulletin Number 41 (Pershing) 68
Bulletin Number 44 (Pershing) 68
Burgess-Wright aircraft 56
Buzancy, France 92

Cadillac automobile 82
calisthenics (training) 55, 58, 143, 145
Cambrai, Battle of (1917) 34
Camden, South Carolina 137
Camp Colt, Pennsylvania 75
Camp Gordon, Georgia 58
Camp Logan, Texas 8, 57, 101
Camp Meade, Maryland 132, 135
Camp Wadsworth, South Carolina 57, 58
canal (Meuse River) 112–16
Cantigny, Battle of (1918) 10, 16, 34
Caproni CA5 aircraft 36
Caquot Type R kite balloon (observation) 77
Carson, Colonel C.C. 82

Index

Casad, Colonel Adam F. 70
Castner, General Joseph C. 107–110, 113–15
Catholic University, Washington, D.C. 21
cavalry 13, 23, 24, 38, 39, 55, 56, 124, 137–39
Cavalry Journal 138
cavalry training 65, 137–39
Center of Artillery Studies Report (Lock) 71
Chaffee, Colonel Adna R., Jr. 135, 139
Chalons-sur-Marne, France 80
Champigneulle, France 96
Chandler, Colonel Charles D. 78
Chatillon-sur-Seine, France 65, 74, 101
Chauchat automatic rifle 31, 74
Chaumont, France 59
chemical factories 130
chemical officers 72
chemical warfare 4, 8, 20–23, 39, 40, 51, 53, 57, 61, 89, 90, 92, 95, 96, 119, 123, 124, 128
Chemical Warfare Service (U.S.) 72, 129, 130, 135
chemical warfare training 65, 71–73, 102, 103, 128–30
Cherpak, Private John 117
Chicago, Illinois 83
chlorine gas 20, 130
chloropicrin gas 20, 130
Choctaw "code talkers" 84
Christie, John Walter 136
Chynoweth, Major Bradford G. 138, 139
Civil War (American) 41, 42, 82
Clamecy, France 66
classroom study 55
Cléry-le-Grand, France 95, 110
Cleveland motorcycle 82
coastal artillery 71, 128
Colt machine gun 58, 73
Colt's Manufacturing Company 23
"combat car" (tank) 139
Combat Instructions (Pershing) 6, 7, 19, 27, 30, 31, 34, 44, 47, 49, 50, 52
Combined Infantry and Cavalry Drill Regulations for Automatic Machine Rifle, Caliber .30, Model of 1909 24
communication 1, 4, 5, 56, 79, 82–84, 87, 93, 96–98, 109, 121, 127, 145, 147
Conflans, France 96
Congress (U.S.) 124, 136
Connecticut Maneuvers, 1912 56
Conner, General Fox 143
Coolidge, President Calvin 141
Cooper, Colonel H.L. 65

corps schools 64–66
Côte Saint Germain, France 96, 115, 116
counter-battery fire (artillery) 22, 51, 90
court martial 134
cover 42, 45, 46, 48, 51, 52, 56, 57, 68, 147
cover (training) 85, 86, 147
covering fire 27, 30, 31, 33, 40, 46, 47, 52, 56, 110, 115, 116, 131
creeping barrage (artillery) 8, 15, 18, 24, 25, 26, 28, 29, 40, 51, 69, 88, 91, 95, 104–07, 110, 126
Croft, General Edward 136
Cunel, France 79, 106, 107, 117
Curtiss JN3 aircraft 36
Curtiss N8 aircraft 36

Davis, Secretary Dwight F. 134, 135, 138
Dawes, General Charles G. 13
decoy targets (artillery) 91
DeHavilland-4 aircraft 36, 77, 141
DeHavilland-9 aircraft 36, 77
Depuy, General William E. 146
destruction fire (artillery) 22
Devers, Lieutenant Jacob L. 127
Dickman, General Joseph T. 96, 128
direct fire (artillery) 56
Dodge automobile 81, 82, 138
Doty, Sergeant Paul A. 119
Douglas A-20 light bomber aircraft 141
Drill Regulations for the 75 French Gun Model 1897, Volume III 18
drum barrage (artillery) 14, 15
Dun, France 90, 92, 97, 112–15
Dun-sur-Meuse, France 109

Eastern Front 15, 48
Edgewood Arsenal, Maryland 130
8th Army (German) 48
80th Division (U.S.) 107
82nd Division (U.S.) 58, 59, 73
88th Aero Squadron (U.S.) 117
Eighty-Eighth Division (German) 110
Eisenhower, General Dwight D. 75, 132–34, 136, 138, 139, 147
élan (fighting spirit) 5, 6, 24, 43, 85, 100, 143, 144
elastic defense (tactics) 26, 99
11th Infantry Regiment (U.S.) 9, 105, 106
Ely, General Hanson E. 10, 91, 107–10, 112, 114, 115, 117–19
Employment of Machine Guns 26, 52
Enfield rifle 32
Engineer Report (Youngberg) 81
engineer training 65, 67, 71, 80, 81, 102

Index

engineers 37–39, 82, 89, 92, 97, 102, 109, 112–14, 118, 119, 124, 135
Épinal, France 103
Erie howitzer factory, New York 127
Excelsior motorcycle 82
Experimental Mechanized Force (U.S.) 135

Farnsworth, General Charles S. 134, 138
Fave River Valley, France 104
The Field Artillery Journal 126
Field Artillery Notes No. 5 17
Field Orders Number 63 (Ely) 108
Field Orders Number 65 (Ely) 109
Field Orders Number 68 (Ely) 112
Field Orders Number 71 (Ely) 115
Field Orders Number 88 (Liggett) 90
Field Orders Number 102 (Liggett) 96
Field Service Regulations 1905 13
Field Service Regulations 1908 24
Field Service Regulations 1923 123, 125, 127, 129, 131, 132, 134, 138, 141–43
Field Service Regulations 1939 125, 127, 136, 139, 142, 144
Field Service Regulations 1941 137
15th Machine-Gun Battalion (U.S.) 9, 103
Fifth Army (German) 48, 99, 100
5th Bavarian Reserve Division (German) 110, 113
V Corps (U.S.) 90, 95, 97
Fifth Division (U.S.) 3, 7–10, 79, 88, 91, 101–21
5th Field Artillery Brigade (U.S.) 9, 102–04, 106
5th Trench Mortar Battery (U.S.) 9
56th Machine-Gun Detachment (German) 114
57-millimeter cannon 134
fighting retreat (German) 98–101
Final Report of Assistant Chief of Staff, G-5 (Harbord) 61
Final Report of the Chief of Artillery, American Expeditionary Forces (Hinds) 69
fire-and-maneuver (tactics) 2, 5, 30, 31, 41, 44–48, 51, 52, 56, 57, 68, 89, 105, 131, 143, 144, 147
fire-and-maneuver training 67, 85, 86, 147
fire at will (rifle) 56
fire discipline (rifle) 46
fire effectiveness (rifle) 32
fire superiority (rifle) 32, 46, 47, 49
firing line (rifle) 49, 56

1st Aero Squadron (U.S.) 36
First Army (U.S.) 8–10, 19, 88–101, 108
1st Cavalry Brigade (U.S.) 124
I Corps (U.S.) 96, 97
First Corps School (U.S.) 65
1st Division (U.S.) 10, 60, 127
1st Gas Regiment (U.S.) 92, 105, 119
1st Tank Brigade (U.S.) 34
Fiske, Colonel Harold B. 61, 68, 71
Flagler, General Clement A. 104
flamethrower 33, 51, 57
flanking maneuver (tactics) 5, 6, 10, 31, 41, 42, 45, 47, 49–53, 105, 130, 143, 144
flanking maneuver training 86
flare (signal) 29, 109, 110
flash ranging (artillery) 19, 80
flexible formations (infantry) 5, 6, 41, 42–44, 52, 143, 145
Foch, Generalissimo Ferdinand 89, 93, 94, 97
Foch's Views on Limited Objectives 94
Fontaines, France 116
footbridge construction 38, 109, 112–15
Ford automobile 82, 118, 138
Ford M1918 light tank 134
Fort Benning, Georgia 124, 142
Fort de Peigney, France 73
Fort Eustis, Virginia 135, 139
Fort Knox, Kentucky 139
Fort Leavenworth, Kansas 42, 60, 63, 67, 82, 102, 143
Fort Logan, Colorado 102
Fort Monmouth, New Jersey 84
Fort Monroe, Virginia 56, 77
Fort Myer, Virginia 138
Fort Omaha, Nebraska 77
Fort Sill, Oklahoma 56, 62, 77, 127
Fort Wise, Texas 77
42nd Division (U.S.) 123
14th Machine-Gun Battalion (U.S.) 9, 103, 108–10
IV Corps (U.S.) 9, 34, 124
Franco-German Border 93, 98, 100, 119, 121
Franco-Prussian War (1870) 43, 45
Frapelle, France 104
French Army 2, 4, 6, 9, 13, 15, 16, 19, 23, 24, 27–33, 36, 39, 43, 45–48, 50–52, 56, 57, 59–61, 63, 75, 86, 87, 90, 100, 101, 103, 104, 131
Fresnes, France 100
friendly fire 4, 15, 16, 18, 20, 29, 33, 46, 97
Fries, General Amos A. 72, 129

Index

frontal assault (infantry) 14, 41–44, 47, 53, 55
Fullerphone 83

garrison training 55
Gas Corps (U.S.) 72, 129
Gas Manual, Part I: Tactical Employment of Gases 20
Gas Manual, Part II: Use of Gas by the Artillery 22, 52
Gas Manual, Part III: Use of Gas by Gas Troops 22
Gas Manual, Part IV: Use of Gas by Infantry 22
gasmask 20–22, 57, 72, 101
Gatling gun 23
General Motors automobile 82
General Orders Number 23-A (Pershing) 89
General Orders Number 29: Instructions for Officers Visiting the French or British Lines or Serving with French, British, or American Units at the Front (Pershing) 61
General Orders Number 35 (Pershing) 67
General Orders Number 70 (Pershing) 78
General Orders Number 77 (Pershing) 80
General Orders Number 81 (Pershing) 78
General Orders Number 82 (Pershing) 73
General Orders Number 91 (Pershing) 73
General Orders Number 131 (Pershing) 80
Geneva Gas Protocol (1925) 129
German Army 2–5, 8, 10, 14, 15, 20, 23, 24, 26, 27, 33, 34, 36–38, 43–45, 47, 48, 51, 52, 54, 56–58, 60, 65, 72, 76, 81, 84, 89–101, 106, 107, 109, 110, 112–14, 116, 117, 119, 121, 130, 131, 136, 137, 142, 145
Gibercy, France 38, 97
Gondrecourt, France 65, 77
Gordon, General Walter H. 104
Great Depression 11, 122, 124–26, 128, 131, 135, 136, 139, 145
grenade 4, 23, 28–31, 35, 39, 40, 45–47, 51, 53, 57–59, 61, 64, 72, 74, 105, 124, 147
grenade training 65–67, 101, 103, 105

Hague Anti-Gas Declaration (1899) 129
hand-to-hand combat 5, 46
harassing fire (artillery) 22
Harbord, General James G. 61, 64, 65, 68

Harley-Davidson motorcycle 82
Heldreth, Sergeant Clyde 105, 117
heliograph 83
"Hello Girls" (telephone operators) 83
Henry, Colonel Guy V., Jr. 135
Hero, General Andrew, Jr. 128
Hero Board (artillery) 128
Hill 252 91
Hill 260 107, 114, 115
Hill 271 107
Hinds, General Ernest 69–71, 97
Hines, General John 90, 106
The History of the Seventy-Ninth Division, AEF 59
Holbrook, General Willard A. 138
horse 5, 13, 23, 26, 33, 38, 39, 56, 70, 84, 90, 93, 97, 98, 108, 113, 114, 128, 128, 135, 138
Hotchkiss M1909 Benét-Mercié machine gun 23
Hotchkiss M1914 machine gun 25, 34, 35, 73, 75
howitzer 16, 69, 95, 102
Huidekoper, Colonel Frederic L. 57, 58, 69, 100

Illinois National Guard 126
Imperial Japan 137
improvisation (tactical) 5–7, 41, 49, 50, 52, 53, 65, 85, 143–45, 147
improvisation training 144, 147
Indian Wars (American) 42, 44, 55
indirect fire (artillery) 96, 119
Infantry Drill Regulations United States Army 1911, Corrected to April 15, 1917 17, 44, 46, 49
Infantry in Battle (Marshall) 126, 130, 131, 144
Infantry Journal 133
Infantry Specialist School Report 74
infiltration (tactics) 5, 8, 10, 41, 47–49, 51–53, 68, 89, 95, 114, 144
infiltration training 85
initiative (tactical) 5, 41, 49, 50, 53, 61, 62, 65, 143, 147
initiative (training) 85, 86, 147
inspection officers 66, 68
Instructions for the Offensive Combat of Small Units 18, 26, 27, 30, 31, 33, 35, 39, 44, 48, 50, 51
Instructions for the Training of Platoons for Offensive Action, 1917 30–32, 47, 50, 51, 85
intelligence (information) 5, 21, 29, 50, 94, 95

Index

interdiction fire (artillery) 22
Iraq (War on Terror) 146, 147
irregular formation (infantry) 6, 42–44, 52
isolationism (diplomatic) 122, 125, 145
Issoudun, France 77
Italo-Ethiopian War (1935) 130, 142

Jametz, France 118, 119
Juvigny, France 118

Kellogg-Briand Pact (1928) 125
Kelly Field, Texas 141
Kriegsmarch (1918) 99
Kriemhilde Line 10, 106, 110, 121
Kromer, General Leon B. 139

Lafayette Flying Corps 77
Lakehurst Proving Ground, New Jersey 129, 130
Langley Field, Virginia 141
Langres, France 65–67, 72, 74
Lanza, Colonel Conrad H. 19, 95, 98, 121, 126
"leap frog" (tactics) 49
"Leavenworth Crowd" 67
Lebel M1886 rifle rounds 25
LeBlanc grenade 29
Legeune, General John A. 95
Le Valdahon, France 59, 71, 102, 103
Lewis, Captain Winford L. 21, 129
Lewisite gas 21, 129, 130
Liggett, General Hunter 8, 9, 86, 88–99, 108
light machine gun 130, 131, 147
limited objective (tactics) 41, 49–51, 53, 89, 94
linear formation (infantry) 49, 53
Liny-devant-Dun, France 109
Livens projector 20, 22, 72
Locke, Colonel M.E. 71
logistics 1, 4, 5, 56, 58, 78, 81, 92, 93, 97, 98, 108, 114, 116, 117, 121, 145
Loison River, France 118, 119
Longuyon, France 96, 118, 119
Loos, Battle of (1915) 20
Lorraine, France 104
Ludendorff, General Erich 27, 98–100
Ludendorff Offensive (1918) 27, 48, 51, 59
Luftwaffe 142
Lynch, General George A. 136
Lyon, France 117

M1 light tank 139
M1 mortar 131
M2 light tank 136
M2 mortar 131
M1885 3.2-inch field gun 14
M1905 bayonet 32
M1917 bayonet 32
MacArthur, General Douglas 123, 124, 135, 139, 145
machine gun 4, 6, 13, 22–27, 30, 31, 34, 35, 38–40, 45, 48, 51–53, 55–59, 61, 64, 78, 86, 90, 92, 93, 95–101, 104–06, 108, 110, 112–14, 116, 117, 119, 122, 123, 125, 130, 133, 135, 136, 139, 142, 145
Machine Gun and Small Arms Activities Report 74
machine-gun officers 73
machine-gun training 65–67, 71, 73, 101, 103, 130–32
Mackensen Wedge (artillery) 15
Malone, General Paul B. 64, 106, 107, 112, 114, 118
manpower shortage 125
Manual for Noncommissioned Officers and Privates of Field Artillery of the Army of the United States, Volume I, Corrected to December 31, 1917 17
Manual for Noncommissioned Officers and Privates of Field Artillery of the Army of the United States, Volume II, Corrected to December 31, 1917 17
Manual for Noncommissioned Officers and Privates of Infantry of the Army of the United States, 1917 32, 41, 46, 47, 50
Manual of the Automatic Rifle (Chauchat), Drill-Combat-Mechanism 31, 51
Manual of the Chief of Platoon of Infantry (French) 15, 24, 28, 30, 33, 43, 45, 46, 48, 50, 74
March, General Peyton C. 62, 123, 140
marching drill 55, 58, 143, 145
marching fire (automatic rifle) 51, 52
Mark II grenade 29
Mark III grenade 29
Mark IV heavy tank 35
Mark V heavy tank 35
Mark VIII heavy tank 134
marksmanship 5, 24, 32, 55, 58, 59, 62, 63, 85, 101, 103, 105, 131, 132, 143, 146
Marshall, General George C. 87, 88, 93–97, 100, 101, 124, 126, 127, 130, 131, 136, 137, 143–45, 147
mass formation (infantry) 14, 41–44, 49, 53
Maxim 08/15 light machine gun 24

170

Index

Maxim machine gun 23, 55
Maxse, General Ivor 43
McAndrew, General James W. 61
McMahon, General John E. 91, 101, 102, 105
McNair, General Lesley J. 127, 144, 145
Medal of Honor 106, 114
Menoher, General Charles T. 140, 141
Metz, France 101
Meuse-Argonne Offensive (1918) 2, 4, 5, 7, 9, 10, 18, 19, 23, 26, 34, 35, 38, 39, 41, 44, 55, 59, 65, 66, 68, 73, 74, 78–80, 82, 86–121, 123, 129, 131, 143, 144
Meuse River, France 8, 10, 95, 97–100, 106, 108–10, 112–17, 119
Mézières, France 92
MG-34 machine gun 130
Middle River, Maryland 130
Mills bomb 29
Milly, France 114, 115
Mitchell, General William L. 37, 95, 140–42
Montmédy, France 96, 118, 119
Morrison, General John F. 62
Morrow, Dwight W. 141
Morrow Board (airpower) 141
Morse Code 83
mortar 4, 13, 20, 22, 23, 27, 28, 39, 40, 51–53, 57–59, 61, 72, 95, 104, 124, 125, 131, 145
mortar training 65–67, 130–32
Moseley, General George Van Horne 81, 82
Motor Transport Corps (U.S.) 81
motorcycle 56, 70, 82, 96, 135, 138
Mouzay, France 117
Mouzon, France 96
Murvaux, France 96, 117
mustard gas 8, 21, 22, 52, 90–92, 130
mutiny 100

National Defense Act of 1916 36, 66
National Defense Act of 1920 125, 129, 133, 138–40
National Guard 42, 56, 58, 146
Native Americans 55
neutrality (diplomatic) 58, 125
neutralizing fire (artillery) 22, 26
A New System of Infantry Tactics (Upton) 42
New York Times 137
Newton mortar 27, 28
Nieuport 28 aircraft 36
night bombing (airpower) 92

9th Infantry Brigade (U.S.) 8, 101, 103, 104, 107–10, 112–14, 117, 118
9th Signal Battalion (U.S.) 9, 117
19th Field Artillery Regiment (U.S.) 9
Noble, Corporal Carl 108, 114, 117, 118
Nolan, General Dennis E. 94, 96, 97
North Africa (World War II) 125, 137, 142, 145
Notes on Grenade Warfare: Compiled from Data Available on February 15, 1917, Army War College 29, 51
Notes on the Employment of Machine Guns 25, 31, 52

observation balloon 77, 78, 92, 105
observation balloon training 65, 67, 77, 78
officer candidate school 66
officer shortage 101, 102
The Official History of the 82nd Division, AEF 59
The Official History of the Fifth Division USA 103, 114, 115, 117
Olsmith, Lieutenant Vernon G. 102, 104
103rd "Lafayette" Squadron (French) 77
105-millimeter howitzer 128
123rd Division (German) 106
131st Infantry Regiment (U.S.) 49
154th Field Artillery Brigade (U.S.) 128
158th Brigade (U.S.) 38
Operation Memorandum Number 119 (Ely) 108
Operations Report Number 27 (Marshall) 94
Operations Report Number 28 (Marshall) 94
Operations Report Number 29 (Marshall) 95
Operations Report Number 32 (Marshall) 96
Operations Report Number 34 (Marshall) 97
Operations Report Number 35 (Marshall) 96
Operations Report Number 38 (Marshall) 101
Ordnance Department (U.S.) 75
Ordnance Report (Casad) 70
Organization and Tactics (Wagner) 16, 24, 35, 42, 46
O'Ryan, General John F. 54, 57, 63, 64, 69, 85, 86

Packard automobile 82
panzer division 136

Index

Passchendaele, Battle of (1917) 15, 19, 47, 51
Patrick, General Mason M. 78, 79, 141
patrol (infantry) 29, 50, 56, 59, 90, 103, 108, 109, 117
Patton, General George S. 34, 74, 75, 87, 132–34, 136, 138, 139, 145, 147
Pearl Harbor Attack (1941) 137
Pershing, General John J. 1–3, 5–8, 10, 13, 16, 18, 19, 23, 26, 27, 30–39, 41, 43–45, 47, 48, 50–54, 59–63, 65–68, 70–73, 75, 76, 78, 80, 81, 85–91, 93–95, 97, 99, 101–04, 106–08, 122, 123, 126, 131–33, 138, 140, 141, 143–47
Philadelphia, Pennsylvania 60
Philippine Insurrection (1899) 10, 44
Phillips, Colonel A.E. 74
phosgene gas 21, 22, 130
phosphorous 21–23, 29
photography 78, 79, 92, 94
pigeon (carrier) 65, 83, 84, 93, 96
Pike, Colonel S.L. 66
Pintheville, France 34, 44
Plan of Radio Liaison (Ely) 109
Plan of Telephone Liaison (Ely) 109
Plunkett, Admiral Charles P. 90, 95
pontoon bridge construction 97, 109, 112, 113, 115, 116
preliminary bombardment (artillery) 14, 40, 88, 92, 95, 104–06, 108, 110, 126
Principles of War 123, 143
prisoners 100, 114, 117, 119
Provisional Machine-Gun Firing Manual, 1917 25
Prussian Army 49
Punitive Expedition into Mexico (1916) 14, 36–38, 44, 81, 138
pursuit aircraft (fighter) 77, 95, 96, 141

quiet sector (training) 69, 103, 104

radio 56, 65, 79, 80, 84, 108–110, 126, 127, 142, 145
railroad construction 38, 81, 82, 89, 92, 93
railway gun (artillery) 16, 90, 95
Raucourt, France 96
reconnaissance 5, 36, 50, 51, 53, 56, 82, 90, 138, 139
Red Army 136, 146
Regia Aeronautica 130
Regular Army (U.S.) 101
Rembercourt, France 106
Remoiville, France 97, 117
Renault FT17 light tank 34, 75, 124, 134

Report of First Army 95
Rhine River, Germany 98, 100, 121
Riaville, France 100
rifle grenade 22, 30, 61, 105, 113
Riga, Battle of (1917) 48
road construction 80, 81, 89, 116
Rochester cannon factory, New York 127
Rockenbach, General Samuel 133
Rohr, Major Wilhelm 48
"rolling kitchen" 8, 9, 117
Romagne, France 107
Root, Secretary Elihu 42, 55
Root Reforms (1899) 42, 55, 67
runner (messenger) 83, 84, 96, 127
rushed training 6, 7, 8, 54, 59, 60, 63, 65, 67–69, 86, 87, 102, 110, 119, 145, 146, 148
Russian Army (Imperial) 48
Russo-Japanese War (1905) 6, 28, 32, 43

Sabine River Valley, Louisiana 137
Saint Die Sector, France 104
Saint Hilaire, France 100
Saint Maur, France 70
Saint Mihiel Offensive (1918) 9, 10, 23, 34, 36, 72, 80, 104–06, 132, 135
Salmson 2A2 aircraft 36
salvo fire (artillery) 17
San Antonio, Texas 141
San Francisco, California 138
San Geronimo, Mexico 39
San Juan Hill, Battle of (1898) 23
Schneider M1897 75-millimeter cannon 16, 18, 65, 93, 102, 108, 109, 124, 128, 129, 136
Schneider 155-millimeter howitzer 16, 65, 102, 118, 119, 128
School Project for the American Expeditionary Forces (Harbord) 65
"scorched earth" strategy 100
Second Balloon Company (U.S.) 105
Second Battle of Ypres (1915) 20
Second Corps School Report (Cooper) 65
Second Corps School (U.S.) 65
Second Division (U.S.) 10, 95, 124, 141
Second Sino-Japanese War (1937) 142
Secret Field Orders Number 73 (Ely) 118
Secret Field Orders Number 75 (Ely) 119
Sedan, France 10, 92, 98, 108
semaphore 65, 82, 93
semiautomatic rifle 124
Services of Supply (U.S.) 81
Seventh Army (French) 103
7th Cavalry Brigade (Mechanized) (U.S.) 139

172

Index

7th Cavalry Brigade (U.S.) 39
7th Engineers Regiment (U.S.) 9, 102, 108, 110, 112–16, 119, 129
Seventy-Fifth Division (German) 119
77th Reserve Division (German) 106
78th Division (U.S.) 106
79th Division (U.S.) 19, 38, 59, 85
Sherman, General William T. 55
shock (tactics) 46, 53
Sibert, General William L. 72
Signal Corps (U.S.) 79, 82, 135
signal lamp 83
Signal Report 83
signal training 65
signalmen 92, 96
VI Corps (U.S.) 140
6th Infantry Regiment (U.S.) 9, 102, 104–06, 112, 113
60th Infantry Regiment (U.S.) 9, 105, 108, 110, 116, 117
61st Infantry Regiment (U.S.) 9, 104, 109, 110, 115
65th Infantry Brigade (U.S.) 34, 44
skirmish line (infantry) 42–44, 52
Small Arms Firing Manual 1913, Corrected to March 15, 1918 32
small-unit tactics 55
Smith, Corporal John H. 104
smoke 21–23, 72, 92, 95
sniping 59
Soissons, Battle of (1918) 16, 25, 34
Somme, Battle of the (1916) 28, 34
Sopwith Camel F1 aircraft 36
Souilly, France 5, 95, 97
sound ranging (artillery) 19, 80
Southwest Pacific (World War II) 125, 142, 145
Soviet Union 136, 146
Spad VII aircraft 36
Spad XIII aircraft 36
Spanish-American War (1898) 14, 23, 32, 42, 44, 46, 55, 57, 77
Spanish Army 55
Spanish Civil War (1936) 142
Special Orders Number 518 (Liggett) 95
Springfield rifle 32, 124, 145
squad (infantry) 45, 105, 147
squad rush (infantry) 49
Squier, George O. 83
standardization (tactics/training) 54, 55, 63, 64, 83, 86, 147
standing barrage (artillery) 14
stateside training 62, 67, 73, 75, 77, 85, 102
Stenay, France 92

Stilwell, General Joseph W. 124, 145
Stimson, Secretary Henry 137
Stokes mortar 20, 22, 27, 51, 113, 124, 131
"storm trooper" (tactics) 48, 51
"strategic missions" (airpower) 141
Summerall, General Charles P. 16, 70, 90, 95, 135
Superior Board (artillery) 128
suppressing fire (tactics) 5, 26, 34, 45, 46, 48, 51–53, 93, 95, 105, 109, 130, 131, 144
suppressing fire training 86
surprise 85, 108, 145
switchboard (telephone) 83

T5 medium tank 136
T-34 tank 136
"tactical missions" (airpower) 141
Tactical Note Number 7, Combat Instructions for Troops of First Army (Pershing) 18, 26, 27, 35, 38, 44, 48, 50, 52
Tailly, France 94
tank 5, 7, 13, 27, 34, 35, 38–40, 51, 57, 61, 83, 87, 89, 94, 95, 105, 106, 122–24, 127, 132, 134, 137–40, 142, 147, 148
Tank Corps (U.S.) 34, 74, 133
"A Tank Discussion" (Eisenhower/Patton) 133
tank training 65, 74–76, 132–37
teargas 29
telegraph 65, 82–84, 92
telephone 56, 65, 77, 82, 84, 92, 109, 117, 127
10th Artillery Brigade (U.S.) 113, 114, 116, 127
10th Infantry Brigade (U.S.) 9, 101, 103–07, 112, 113, 115–18
tetrachloride smoke 21
Thinte River, France 38
3rd Artillery Brigade (U.S.) 108, 110, 113, 118, 129
3rd Attack Group (U.S.) 141
III Corps (U.S.) 90, 96, 97, 106, 135
3rd Infantry Brigade (U.S.) 10
3rd Tank Brigade (U.S.) 95
13th Machine-Gun Battalion (U.S.) 9, 103, 118
Thirty-First Division (German) 106
XXXIII Corps (French) 103
33rd Division (U.S.) 34, 44, 49, 57, 58, 69, 100
36th Division (U.S.) 84
37-millimeter cannon 33–35, 40, 52, 57, 59, 61, 75, 93, 109, 124, 136

173

Index

37-millimeter cannon training 65–67
311th Machine-Gun Battalion (U.S.) 19
304th Engineer Regiment (U.S.) 38
326th Tank Battalion (U.S.) 135
332nd Infantry Regiment (German) 106
Toul Sector, France 103
tractor 69, 70, 82, 128
Training Area Thirteen, France 102
training maneuvers 55, 122–125, 127, 135–38, 141–45, 147
training officers 57, 59–61, 68, 77, 101–04
trench artillery 27
trench raid 28, 29, 45, 59, 90, 103
trench raid training 102
trench warfare (tactics) 1, 2, 16, 24, 27, 43, 45, 60, 63, 86, 122, 143
trench warfare training 102, 103
12th Aero Squadron (U.S.) 105
20th Field Artillery Regiment (U.S.) 9
28th Infantry Brigade (U.S.) 10
Twenty-First Division (French) 104
21st Field Artillery Regiment (U.S.) 9
27th Division (U.S.) 54, 57, 58, 63, 64, 69, 73, 84–86
284th Aero Squadron (French) 108

unconditional surrender 97
unconventional warfare (guerrilla) 55, 101, 146, 148
Upton, Colonel Emory 42, 44, 49, 50, 53, 86

Van Voorhis, Colonel Daniel 135
Verdun, Battle of (1916) 15, 48, 51
Very flare pistol 29
Vickers machine gun 25, 73
Villa, Francisco "Pancho" 14, 36
Villers-devant-Dun, France 94
Vimy Ridge, Battle of (1917) 47
Vincennes, France 69
Vivien-Bessière rifle grenade 30
volley fire (rifle) 17, 46, 56, 57
volley sweeping fire (artillery) 17

von Baden, Prince Max 100
von der Marwitz, General Georg 99
von Hindenburg, General Paul 98–100
von Hutier, General Oskar 48
von Mackensen, General August 15
Vosges Mountains, France 103

Wagner, General Arthur L. 16, 17, 24, 42, 46
War College (U.S.) 42, 95
War Department (U.S.) 16, 18, 24, 26, 32, 36, 37, 39, 47, 57–60, 62, 63, 67, 72, 75, 101, 123, 125, 127, 130, 136, 137, 139, 140
War Department Training Regulations Number 10–15 123, 143
war games 67
War on Terror 146, 148
The War with Germany: A Statistical Summary (Ayers) 70
Ward, Lieutenant Orlando 127
Washington, D.C. 138
Watertown arsenal, Massachusetts 127
Watervliet arsenal, New York 127
weapon shortage 54, 56–58, 69, 70, 73–78, 81, 82, 87, 90, 93, 94, 97, 101, 119, 127, 128, 137, 145, 146
weather reporting 82
West Point, New York 10, 60, 63, 123
Western Electric 83
Westervelt, General William I. 128
Westervelt Board (artillery) 128
White automobile 82
Wilhelm II, Kaiser 101
Wilson, President Woodrow 58
wiretapping 109
Woodfill, Lieutenant Samuel 106
World War II 3, 10, 11, 41, 52, 79, 122, 124, 125, 127, 128, 130, 131, 136, 137, 142, 145, 148

Youngberg, Colonel G.A. 81

Zinn, Captain Frederick W. 77